ANTHROPOLOGICAL PAPERS OF
THE UNIVERSITY OF ARIZONA
NUMBER 25

# IRRIGATION'S IMPACT ON SOCIETY

editors | THEODORE E. DOWNING and McGUIRE GIBSON

collaborating authors |
Robert McC. Adams
Theodore E. Downing
Ian Farrington
Chris Field
Martin M. Fogel
McGuire Gibson
Eva Hunt
Robert C. Hunt
Wayne Kappel
Susan H. Lees
M. Edward Moseley
James A. Neely
Robert McC. Netting
J. E. Spencer
Brian Spooner
R. Gwinn Vivian

THE UNIVERSITY OF ARIZONA PRESS
TUCSON, ARIZONA 1974

THE UNIVERSITY OF ARIZONA PRESS


Manufactured in the U.S.A.

ISBN-0-8165-0419-9
L. C. No. 74-15602

# CONTENTS

ILLUSTRATIONS

# PREFACE

The relationship of irrigation to society has been a subject of considerable investigation since Wittfogel formulated his Hydraulic Society, in which irrigation necessitates centralized control, bureaucracy, and other features of civilization. Other scholars, through research in specific geographical areas or through symposia that took a comparative approach, rejected the formulation totally or in part, modified it, or proposed alternative theories.*

The increasing popularity of ecological and ethnoscientific studies in anthropology has rekindled interest in the irrigation society equation. Earlier investigations sought the social impact of this technological change; environment formed a backdrop, rather than an intimately related and changing part of the total ecological process. Recent approaches, including the papers in this volume, differ in emphasis. They proceed from a consideration of natural factors, including population, to view irrigation as yet another means for adapting to a specific environment. A related interest in society's effect on irrigation gives the contributions a coloring that makes the title of the book, and the symposium on which it was based, seem somewhat anachronistic. The book might more precisely be titled "Society's Impact on Irrigation," or "The Impact of Society and Irrigation on One Another."

Initially, when we decided to organize a symposium at the 1972 meeting of the Southwestern Anthropological Association in Long Beach, California, we hoped there would be enough interest to make possible an afternoon session. A few phone calls made it clear that we were underestimating our subject. Literally dozens of scholars in several disciplines were studying the relation of irrigation to society. Participants were eager to come from as far away as Boston, New York, and Philadelphia, although we were unable to offer any compensation. Obviously irrigation was still a "hot" issue.

On April Fool's Day, the contributors spent the entire day discussing this serious subject. All the papers except one are reproduced in this book. Thomas F. Glick, a historian, has published his lengthy essay, *The Old World Background of the Irrigation System of San Antonio, Texas,* as a separate monograph (El Paso: Texas Western Press, 1972).

As discussants, we had Martin Fogel, a hydrologist and engineer, and Chris Field, a geographer. Lewis R. Binford, who was to serve as our archaeological commentator, was unable to attend due to unforeseen circumstances.

Given the diversity of disciplines of the participants (archaeology, social anthropology, geography, and history) and the diffuse sources drawn upon for evidence and theories (adding agronomy, hydrology, and systems engineering), one might expect widely divergent results in the papers. However, despite this eclecticism, which is characteristic of social research, and despite the fact that we consider irrigation systems from different parts of the world and in ancient and modern times, there are several shared themes that tie the book together.

Notions of cycles expressed by Adams, Gibson, and Lees were arrived at independently either through a consideration of environment in a common geographical area (Adams, Gibson), or as part of a shared theoretical viewpoint (Gibson, Lees).

In most papers (esp. the Hunts, Downing, Netting, and Spencer), emphasis is placed upon discovering the relevant elements of the physical irrigation system.

Comparable water-control solutions were discovered for two unrelated cultures that had similar problems in water scarcity on generally comparable terrain (Neely for ancient Iran and Vivian for ancient southwestern U.S.).

Underlying many of these presentations is a concern with the problem of relating the complexity of social organization to the complexity of water-control problems. Evaluating that relationship requires some measurements of the complexity of both the physical irrigation system and social organization. It is clear that the gross size of an irrigation system need not be a measure of complexity of the physical system or of the social system that controls it; the size of the irrigation system needs more careful consideration. "Large" for the Deh Luran Plain (Neely) would be considered "small" in acreage or as

*See Eva and Robert Hunt's paper in this book for an excellent review of the history of research on the problem.

systems when compared to the American Southwest (Vivian) or modern Mexico (Lees, Downing). Even where canal lengths are comparable, they may vary in function and effect. For instance, one would have a very different set of managerial problems in dealing with a major river like the Euphrates that carries too much water at the wrong time, than one would have in ancient Mexico, where massive canals were built to carry a trickle of water for long distances. Neely's Iranian case, in which a relatively small irrigated area involved the construction of expensive technological devices, is more comparable to other areas described in this book than to its neighbor, Mesopotamia.

According to Spooner, the investment involved in a specific irrigation scheme, whether large or small in area, must be taken into account when scaling complexity. A relatively small system (areally) may represent a tremendous investment in time, manpower, money, and centralized organization, while an extensive system may require only a few days of annual, volunteer, tribally organized labor (Adams, Gibson). Piecemeal construction by relatively low-order organizations, such as neighbors, kinsmen, or tribes can result in large irrigation systems, but represent minor expenditures (Netting, Spencer, Farrington, Adams, Gibson). Piecemeal construction and the element of time are emphasized by Spencer: small social units, working steadily through time, build up an investment that is, in actuality, very large. However, with this incremental investment, there may result social arrangements no more complex than are needed to make possible the individual units of construction, maintenance, and the like.

Water is used differently in Spencer's case than in the others described in this book. Here, the water is more a nutrient carrier, recycled as often as possible, rather than merely a source of moisture. In this particular situation, no problem of scarcity exists, and it might be argued by some contributors that, lacking the pressure, no complex social organization is necessary. The question of scarcity becomes critical in evaluating the Wittfogel hypothesis. Most of the archaeological papers in this volume fall on the anti-Wittfogelian side of the issue and argue that political organization precedes complex irrigation (Adams, Gibson, Neely, Vivian, Farrington, Moseley). Their position finds support in Kappel's comparative analysis of modern irrigation situations. Lees, the Hunts, and Spooner, on the other hand, present evidence that irrigation *does* affect social organizations and bring about stratification *if* there is a scarcity of water, or *if* conditions exist that, in effect, restrict access to water or land. Downing presents the disturbing possibility that "water scarcity" may be crop-specific and emphasizes the importance of considering alternative water sources such as rainfall.

Scheduling of water for specific cultigens, especially when water is limited, is very much a part of Wittfogel's hypothesis. The need for precise timing of waterings, in his view, requires centralized time-reckoning, astronomy, and record-keeping. However, among our papers is Netting's intriguing Alpine case, in which intricate scheduling and partition of water are accomplished not only without centralized authority, but without the conscious understanding of how the system works at all.

There does seem to be general agreement among the contributors, even those who argue for the priority of social complexity over irrigation, that once having set up irrigation agriculture and having tied subsistence to it, there is a tendency toward centralization. The key to this centralization may well be scarcity, as proposed by the Hunts, Downing, and Lees. However, the scarcity may not be merely one of water. The vital factor seems to be stress on resources or a limitation of access to them. Population pressure may be the cause of stress in one situation, while a drop in population, which reduces manpower, may be crucial in another. The inability of local groups to furnish necessary capital for an irrigation system capable of competing economically (as in Spooner's case) may be the trigger at times. Likewise, the need to invest in non-agricultural goods may cause governments to assert control. Scarcity of land through salting or alkali action, especially when combined with population rise or any of a number of other factors, may be the key in a given time and place.

Emphasis on individual situations, with their own particular stresses, is a major concern of most of the contributors. There seems to be a feeling that it is time to return to the data, do detailed studies of individual irrigation systems, work out the technical details of engineering, and assess how the system relates to the natural and social environments. A wide range of factors – such as water source, volume, velocity, quality, scheduling, and application – have been introduced for detailed consideration. There seems to be a need for a typology, or typologies, of irrigation systems. For comparative purposes, we must begin to use technical terms, such as "floodwater irrigation," with consistency and with the same meaning as is employed by hydrologists. There is a great need for mapping present-day systems in areas unaffected by modern technology, and relating to them the social systems concerned with them. Having done that, it should be possible to determine whether a social organization receives its essential structure from factors outside irrigation needs. Social anthropologists and archaeologists must delve more deeply into the fields of the geographer and hydrologist, while the latter should become more anthropological.

As a closing note, we would like to thank all participants in the symposium for their continued enthusiasm and interest. All sacrificed time, other duties, and money to present their papers and revise them for the book. Deadlines were admittedly severe, but were met with good spirits; editorial changes were suffered with good grace. Some readers may find the elimination or translation of commonly used local terms (e.g. *shaykh*, *mitimaes*) a bit annoying, but we operated upon the principle that people from different disciplines, knowing nothing of the language in a given area, should be able to read without a language dictionary.

Sue Sogge, the Department of Anthropology editor, deserves praise for her editorial assistance, as does S. C. Newmark for preparing the index and Robert Williams for his assistance. We would especially like to thank Raymond H. Thompson, chairman of the Department of Anthropology, University of Arizona, for his encouragement, advice, and support.

Theodore E. Downing
McGuire Gibson

CHAPTER 1

# HISTORIC PATTERNS OF MESOPOTAMIAN IRRIGATION AGRICULTURE

Robert McC. Adams
*Department of Anthropology, University of Chicago*

Historically attested systems of irrigation in Mesopotamia can be grouped into a number of contrasting configurations. Great differences in scale and complexity are most immediately apparent, but with these occur corresponding variations in the social and institutional arrangements by which irrigation agriculture was conducted. However, since there was a demanding, highly persistent set of ecological constraints imposed on all such systems, important elements of continuity appear in the traditional agricultural technology for more than six millennia. Shaped also by political and other variables, the historic outcome has involved many episodes of an unstable alternation between patterns, as well as an underlying regularity in the succession of developmental stages. To understand the elements of both change and continuity, it is useful at the outset to consider the natural conditions on which all Mesopotamian irrigation systems of the past have depended.

Although low but varying rainfall plays an appreciable part in the success of an individual harvest, techniques of irrigation have always been indispensable to agriculture. Until recent times, primary emphasis has always been placed not on pumps or other lifting devices, but on gravity-flow canals. Obtaining their flow from the major water courses on elevated, natural levees, gravity-flow canals irrigate lands situated on the levee backslopes and depression margins, as well as further downstream. Cereal crops, principally wheat and barley, require four or more liberal applications of water during the winter growing season, with alternate years in fallow. Low water levels in the rivers have always limited cultivation during the intensely hot and dry summers and have prevailingly limited cultivation to as little as one-tenth or less of the winter acreage. High water levels, often accompanied by destructive flooding, are ill timed for maximizing cultivation and occur only at the end of the growing season in the spring (Wirth, 1962).

The fallow system, often regarded by specialists in temperate agriculture as wasteful, in fact, serves several vital functions. Deep-rooted, perennial, nitrogenous weeds survive the traditionally shallow plowing and provide a natural source of both firewood and fertilizer. Equally important, their virgorous growth during years out of cultivation reduces the threat of salinization by lowering the saline groundwater table. Since the supply of potentially arable land far exceeds the amount that can be cultivated with the available supply of water, there is no inducement to intensify cultivation of limited areas. Hence fallow fields, stubble, seasonal marshes, and semi-arid steppe provide permanent sources of fodder for herds of sheep, goats, cattle and other animals that are complementary to rather than competitive with the uses of land for cultivation. Similar complementary relationships can be traced between the contributions of cultivated plants and the products of animal husbandry to a balanced diet, and between the brief but heavy labor requirements of the harvest and the availability of a supplementary labor supply among herdsmen. In short, the natural conditions of the area impose extensive rather than intensive systems of irrigation agriculture. In addition, these conditions place a premium on institutional arrangements to augment the supply of agricultural labor, whether through cooperation, reciprocity, or corvée. They strongly favor not monoculture but a balance between cultivation and animal husbandry.

It is crucial to recognize the deep, chronic instabilities of agricultural subsistence under these circumstances. Major stream channels are prone not only to flooding but to destructive shifts in course and variations of flow that repeatedly jeopardize whole areas. Smaller canals and natural distributaries require organized, annual removal of the silt that accumulates in their beds. If disturbed political conditions prevent such efforts, these smaller channels may choke up quickly and in large numbers, vastly intensifying a crisis. Uncertainties as to the supply of water reinforce a tendency to over-irrigate whenever water is available, intensifying destructive competition between upstream and downstream consumers and aggravating the problem of soil salinization. Combined with other hazards such as crop blights and insect infestations, the effect of these and similar features is periodically to force both local and regional impoverishment or even abandonment. Myths about the productivity and reliability of irrigation agriculture notwithstanding, plentiful, assured yields in return for inputs of labor at the disposal of the individual cultivator simply do not occur. To the small, individual farmer or husbandman, it is a system of low return and high risk. Hence arrangements to pool resources and spread risks play a vital adaptive role, encompassing social groups beyond not only the nuclear family but even the local community. Tribal, temple, state, or absentee landlord tenure may be viable alternatives in different historic circumstances, but in any case there are disproportionate advantages attached to large landholding units (Fernea, 1970).

Turning from general conditions that confronted all agriculturalists, we may ask how they were met in different ways in successive epochs. The earliest configuration known in any detail dates from the latter half of the fourth millennium. Only a few centuries prior to its transformation into an overwhelmingly urban landscape, southern Mesopotamia was comprised of a few isolated ceremonial centers and a large number of villages and small towns.

From the interpretation of patterns of distribution obtained through archaeological surface reconnaissance, it is clear that settlement followed the numerous small, meandering stream distributaries found on any unmodified alluvial plain. Lengthy, branching canal networks are not in evidence, and cultivation seemingly was confined to narrow bands along natural levee backslopes and to favorably situated margins of seasonally filled depressions. There are some identifiable territorial units consisting of small villages grouped around a larger site, the latter perhaps having adjoined and furnished protection to a brushwork weir serving a number of small canal offtakes. Sites for the most part were scattered more loosely, however, and fail to suggest even this fairly rudimentary degree of centralized control over the irrigation system.

An essentially similar pattern reappeared at many later periods and predominated during intervals when strong, centralized political controls deteriorated. Such was the case in particular during the late second and early first millennia B.C., and from late Abbasid times until the early twentieth century A.D. Since rural conditions tend to be poorly described in indigenous urban documents, this pattern can be best understood from travellers' accounts from the late eighteenth and nineteenth centuries as well as from more recent ethnographic reports (Fernea 1970; Adams and Nissen 1972).

Although population levels were low in relation to water supplies theoretically available for the alluvium as a whole, these sources depict a situation of localized but chronic water shortages. Irrigation was in the hands of tribal units of varying scale and was extended and maintained in small increments without overall plan or recourse to a specialized "hydraulic" bureaucracy employing means of coercion. Movement of villages and whole tribal domains was frequent, partly in response to local conditions and partly to rapidly shifting balances of military power that were not infrequently precipitated by Ottoman authorities in order to promote their own ascendancy. Substantial investment in large scale, permanent irrigation works was obviously impractical under such circumstances, and was further discouraged by the prevailing instability in the distribution of flow among the major natural channels on which any canal system had to depend.

However, the constant flux within the system does not imply that it was only briefly or marginally effective as an adaptation to the natural conditions set forth earlier. Tribal constituencies normally included sections composed of semi-nomadic elements as well as settled cultivators. This diversity not only assisted in meeting both groups' dietary needs, but preserved their joint option to swing decisively toward or away from sedentary life as conditions required. Tribal sub-groups also functioned as organizational units and for risk- and resource-sharing purposes, in warfare, canal construction, and the everyday tasks of agricultural life. Leadership was, to be sure, markedly non-managerial in character. Instead there were strong reciprocal ties between shaykhs and their tribal followers that depended on such charismatic qualities of the former as openhanded generosity, fairness in adjudicating disputes, and bravery in warfare.

The foregoing pattern obviously is most consistent with relatively low population levels, and runs counter to any but a weak, decentralized political system. It is also most consistent with a society that may exhibit considerable social ranking but that lacks clear principles of class stratification. Yet by the early third millennium, both a political and a class structure appeared as outgrowths of this pattern, closely accompanying what has justifiably been called an "Urban Revolution." Clearly, there must be a relationship between at least some features of the pattern and the developments that followed.

Some scholars have argued that dependence on irrigation played such a role, engendering the growth of a bureaucracy whose agromanagerial responsibilities placed them in monopolistic control of a despotic, hydraulic state (Wittfogel 1957). At least for Mesopotamia, the available evidence strongly argues otherwise. Large-scale irrigation systems did not exist at the time states made their first appearance, and nineteenth-century and recent practices in the area make clear that there is absolutely no requirement for a bureaucracy of any sort to construct and maintain irrigation networks even larger and more complex than those that are known from the late fourth and early third millennia. In addition, textual sources of the mid-third millennium – the earliest to shed substantial light on the problem – suggest a pluralistic basis for agricultural management even after the consolidation of political power in dynastic city-states had continued for some centuries, and fail completely to disclose a rigid state superstructure rooted in its control of hydraulic works.

Several alternative features of the decentralized, early (but repeatedly recurring) agricultural pattern seem much more likely to have contributed to the formation of the state. Principal among them is the chronic flux and local instability in the system itself. This instability would have conferred important advantages upon enlarged, organized population clusters, including greater military strength for competition with other communities over water and other scarce resources, as well as better-protected, larger food reserves with which to meet emergencies. Viewing the growth of the state as the political expression of a society increasingly stratified in terms of the differentiated access of its members to productive resources, powerful stimuli also can be discerned at the intra-community level. Prominent among the latter are local differences in land or labor productivity, stemming from the availability of adequate irrigation water or phenomena like salinization. Finally, it has been suggested that animal husbandry also can be a force for economic differentiation, consistently generating a flow of both wealthy entrepreneurs and impoverished laborers into the ranks of the sedentary cultivators (Barth 1961). There are, in short, a number of potentially more productive links to explore between early Mesopotamian agriculture and the emergence of the state than that which is suggested by persistent obsessions with hydraulically-based despotism.

Stable enclaves of villages and towns continued for much longer in the northern part of the alluvium, but the classic Sumerian city-states rapidly emerged in the south during the early third millennium. In a setting of increased emphasis on militarism, it appears from recent surveys around Uruk and Umma that the predominantly rural population overwhelmingly sought out the protection of city walls, within the span of a few centuries. This massive transfer must have involved many correspondingly important changes in the irrigation system. Hence, although some of the changes remain as yet quite hypothetical, we may speak of the emergence of a second major configuration.

Most striking among its features was the abandonment of large tracts of formerly settled, arable land that had been watered by numerous, small natural channels. Whether or not intended, an effect of the depopulation of these areas was the formation of a series of buffer zones between the major city-states, which were in increasingly bitter contention with one another. Equally evident in the results of surface reconnaissance is the consolidation of the stream pattern, diverting water from areas no longer cultivated in order to provide the larger, better-assured supplies to limited areas on which city growth depended. One aspect of this process was the development of riverine transport systems to supply the concentrated urban populations. It is noteworthy that the objective of urban supply, rather than irrigation, is stressed in the few early royal inscriptions dealing with watercourse maintenance, and that the principal terminological distinction is between navigable and non-navigable channels rather than between rivers and artificially constructed canals. Since the urban populace remained overwhelmingly agricultural in occupation, it can be assumed that some rearrangement and intensification of irrigation regimes was undertaken to reduce normal commuting distance to the fields.

By the mid-third millennium, there is a reference by one of the rulers of Lagash to his having diverted Tigris water southward into portions of Sumer previously watered only by the Euphrates. This is the earliest significant attestation of the use of that larger and more destructive watercourse for purposes that must have included irrigation (Jacobsen 1960). Presumably it reflects both an intensifying need for alternative water supplies less jeopardized by the diversions of city-states farther upstream, and increased organizational and technical capacity to secure them. In the late third and early second millennia, there are additional, if still sporadic, references to fairly large-scale canal and dam construction. Textual sources also indicate a comprehensive administrative concern for de-silting and other maintenance and for the construction of protective dikes and small reservoirs (Sauren 1966). There is little to suggest, however, that irrigation at this time depended in any fundamental way on the coercive powers of a monopoly bureaucracy. Recently published texts from the kingdom of Larsa, for example, depict the mobilizing of large numbers of men for canal construction not through repressive corvées but as wage laborers recruited and employed by private contractors (Walters 1970).

It should be stressed that major canal-building efforts were not commonly a part of the locally nucleated irrigation systems that were organized by and around city-states. How successful the rare exceptions may have been initially is difficult to judge from diffuse, self-serving royal claims. But in any case there are a variety of indirect indications that large, artificially constructed canals remained in operation only briefly, generally not outlasting the reigns in which they were initiated. Moreover, the topography as a whole remained little altered by such artificially constructed components as there were in the irrigation systems. There are references, for example, to the employment of sinuous, meandering canals for irrigation, whose description is indistinguishable from that of natural alluvial streams. As late as the Neo-Babylonian period, carefully compiled documents dealing with centrally managed date plantations around Uruk made clear that cultivated areas were not consolidated through drainage into extensive, homogeneous blocks, but instead were interspersed with both seasonal and permanent swamps (Cocquerillat 1968). Unfortunately, this valuable cadastral archive has few parallels. Detailed, direct descriptions of a mundane, rural phenomenon like irrigation were beneath the notice of most urban-oriented chroniclers until the threshold of the modern era.

The third major configuration is that of large scale, comprehensively planned, state-controlled irrigation. Such systems reached their furthest development throughout the Mesopotamian alluvium only in the Sassanian period (A.D. 226-640), but many of their attendant features appeared earlier, and from time to time, on a more restricted geographic basis. This earlier appearance was particularly true during the Old Babylonian (early second millennium) and Neo-Babylonian periods, when there is evidence not only that Babylon itself grew to unprecedented size, but also that the surrounding region was intensively settled. Elements still persist of an extensive scheme of canalization linking the Tigris and the Euphrates in the northern part of the plain, a plan dependent on calculations of differences in their bed levels that probably first became operational in the Neo-Babylonian period. And certainly by late Parthian times, during the first and second centuries of the Christian Era, canals of the size and morphology associated with predominantly artificial rather than natural systems had become widespread.

As epitomized by developments during the Sassanian period, this configuration involved an ambitious expansion and re-shaping of the water supply that for the first time transformed a naturally variegated landscape into a continuously agricultural one. Population rose steeply in the countryside as well as in the cities. Quite possibly this led for the first time to a condition of general water shortage, rather than local shortages based on uneven distribution. Some of the more massive irrigation undertakings, such as the gigantic Nahrwan canal east of modern Baghdad, concentrated on previously underutilized sources like the Tigris, and hence permitted an unprecedented extension of the agricultural area (Adams 1965). Also unprecedented was the emphasis on long, complex, branching systems of secondary and tertiary canals. The straightness and uniformity of these systems is immediately apparent in modern air photographs. Together with uniformity in the baked brick construction of the associated canal offtakes and headworks, this evidence confirms not only the essential artificiality of the watercourses involved but also their centrally planned and administered character.

A close and vital relationship is evident between large-scale irrigation and Sassanian imperial policy. Masses of war captives were employed in primary construction, and many of the newly planned systems were intended for resettled populations. Individual projects in many cases were designed to supply newly founded royal cities and their agricultural hinterlands which enhanced the centralized authority of the dynasty, as well as to offset the strength of the landed nobility in other districts. Fiscal reforms in the later Sassanian period underline the dependence of the growing governmental bureaucracy on agricultural tax income that stemmed largely from the Mesopotamian alluvium. This was a regime, in short, to which the conception of a hydraulically based oriental despotism seems in most respects applicable. Yet it is also worth noting that the regime's origins, as well as its enduring sources of military and political power, were to be found not in the alluvium but in the mountainous highlands of neighboring Iran.

While centralized governmental control and a massive infusion of imperial capital multiplied the productive forces of the countryside, in the long run they were also responsible for decisive shortcomings in the new irrigation systems. Truly large-scale irrigation works depended on state initiatives not only for their construction but for their subsequent maintenance. In the event of a political or military crisis that depleted the state's resources or diverted them elsewhere, the regions served by the new canal systems lacked capital and organization to supply what was missing. In two respects, moreover, the problem was even more acute than the increased size of the canals and the areas they served would seem to indicate. Lengthy, branching canal systems could only be laid out with inadequate slopes for the additional flow they introduced; this led not only to the accumulation of silt in much longer canal beds, but also to a much more rapid rate of accumulation. At the same time, levees formed by the new dendritic systems interrupted natural patterns of drainage. Particularly with the additional flow that the canals made possible, this interruption of the natural drainage patterns had the effect of hastening loss of arable land to salinity by disastrously raising groundwater levels.

Briefly then, expanded agricultural output was attained not only at the cost of greater ecological fragility, but at a cost of greater vulnerability to political and military disturbances. Deep, cyclical fluctuations, with agricultural expansion, prosperity, and rising population followed by deterioration and famine, began during the Sassanian period and intensified thereafter. And to their ecological causes may be added a number of political and economic weaknesses that became progressively more serious and general after the first few centuries of Islam. The rising influence of successive cliques of Turkish bodyguards led to a corresponding decline in the role of the old imperial bureaucracy. This declining bureaucracy weakened the cadres concerned with irrigation management and shifted their objectives away from long-term husbanding of resources toward short-term personal gain. Tax farming, speculation in land, corruption, and even the occasional outright destruction of vulnerable irrigation facilities during uprisings or military intrigues all undermined the system further. With the military expansion of an earlier period having given way to political fragmentation, a shrunken, impoverished agricultural population was left to meet the voracious and inelastic demands of the capital. Large-scale, centrally managed irrigation could not long survive in the face of pressures like these; it had essentially disappeared in much of Mesopotamia even before the *coup de grâce* delivered by the Mongol conquest in the mid-thirteenth century. What replaced it, at catastrophically lower levels of population and economic interchange, was the first, simplest, and most resilient of configurations described above, which had been antecedent to cities and hence would survive their destruction.

In recent times a fourth, neo-technic configuration has begun to appear. Diesel pumps, now having been employed in increasing numbers for more than a century, were perhaps its earliest feature. Reinforced concrete storage dams, weirs, and flood-control works are later but increasingly significant additions. Comprehensive drainage schemes have been added in certain areas within the past fifteen years or so, and the use of agricultural machinery, chemical fertilizers, hybrid seed, and more intensive crop rotational systems all are increasing rapidly. The question nevertheless remains whether the physical resources and limitations of the region in the end will justify large-scale investment in the intensification of agriculture. An extensive system, whose cultivated areas and balance with animal husbandry have been continually adjusted as salinity and other conditions make necessary, has repeatedly confirmed its viability over a span of more than six millennia. It would require not an act of judgment but of faith to proclaim, on the basis of the very brief, recent experience to date, that this oldest and most flexible of the agricultural configurations that Mesopotamia has known will shortly disappear without a trace.

## REFERENCES


ADAMS, Robert McC.
1965 *Land Behind Bagdad: A History of Settlement on the Diyala Plains.* Chicago: Univ. of Chicago Press.

ADAMS, Robert McC. and Hans J. Nissen
1972 *The Uruk Countryside.* Chicago: Univ. of Chicago Press.

BARTH, Fredrik
1961 *Nomads of South Persia: The Basseri Tribe of the Khamseh Confederacy.* Oslo and London: Oslo Univ. Press and Allen & Unwin.

COCQUERILLAT, Denise
1968 *Palmeraies et cultures de l'Eanna d'Uruk* (520-59). Auggrabungen der Deutschen Forschungsgemeinschaft in Uruk-Warka, Bd. 8. Berlin, Gebr. Mann Verlag.

FERNEA, Robert A.
1970 *Shaykh and Effendi: Changing Patterns of Authority among the El Shabana of Southern Iraq.* Cambridge, Massachusetts: Harvard Univ. Press.

JACOBSEN, Thorkild
1960 The Waters of Ur. *Iraq* 22:174-85.

SAUREN, Herbert
1966 Topographie der Provinz Umma nach den Urkunden der Zeit der III. Dynastie von Ur: Teil I, Kanale und Bewässerungsanlagen. Ph.D. diss., Ruprecht-Karl-Universitat, Heidelberg.

WALTERS, Stanley D.
1970 *Water for Larsa: An Old Babylonian Archive Dealing with Irrigation.* New Haven: Yale Univ. Press.

WIRTH, Eugen
1962 Agrargeographie des Irak. *Hamburger Geographische Studien* 13.

WITTFOGEL, Karl A.
1957 *Oriental Despotism: A Comparative Study of Total Power.* New Haven: Yale Univ. Press.

CHAPTER 2

# VIOLATION OF FALLOW AND ENGINEERED DISASTER IN MESOPOTAMIAN CIVILIZATION

McGuire Gibson
*Oriental Institute, University of Chicago*

Descriptions of Mesopotamian civilization routinely include such statements as the following:

> ...a particularly close relationship exists between the flourishing of irrigation agriculture and the existence of a stable and vigorous central government. It is when government controls weaken and disturbed conditions come to prevail that the great disastrous abandonments of land take place (Jacobsen 1958a: 85).

In this paper, I will suggest that, on the contrary, in Mesopotamia the intervention of state government has tended to weaken and ultimately destroy the agricultural basis of the country.

The key to the situation is a clear understanding of the natural factors involved in living on the alluvial plain. Most of southern Mesopotamia has insufficient rainfall to permit dry farming. To farm, one must irrigate. Tidal irrigation, in which tides in the Persian Gulf push fresh water back up the Shatt-al-Arab for more than a hundred miles, may be the earliest type of water use, but the area involved south of Basra has not been investigated systematically and cannot be assessed. Lift irrigation, involving pumps or other devices, played no appreciable part in Mesopotamian agriculture until the more recent historical periods. Even for modern times, it can be argued that this type of water use has been supplemental to flow irrigation.

For the greatest part of flow irrigation, the nature of the Tigris and Euphrates makes it possible and desirable to tap only the latter river in the alluvial plain. The Tigris is unpredictable, fast-running, and cuts into the alluvium to a depth of three meters, making it impossible to use its water north of modern Kut without the aid of pumps or lifting devices (Fig. 2.1). Even in its southern stretches, the Tigris seems always to have played a secondary role. The Euphrates, because it travels a much greater distance, loses about 40 percent of its water through evaporation in the Syrian Desert, must carry a greater load of silt, and arrives on the plain at a much slower speed. The river has a tendency to run above the plain, rather than cut into it, and builds up high levees. In antiquity (Fig. 2.2), the river was a braided stream, dividing into as many as five channels in the area of Sippar (Adams 1958). Through time, the natural channels have been modified by man into artificial courses. Today, there are only two branches of the river, the Hindiyah and the Hilla.

Natural and human factors, causing a greater deposition of silt in the western part of the plain, have resulted in a drainage pattern that slopes from the northwest to the southeast (Fig. 2.1). Ancient and modern canals conform to this drainage system.

To carry out flow irrigation from the Euphrates is relatively simple, since, due to levee formation, the bed of the river lies above the level of the plain. The consequences of tapping the river are, however, complex. There are great variations in the amount of water carried by the river from year to year and from season to season. Although one automatically thinks of the problem for this semi-arid area as one of scarcity, the engineering is as much concerned with getting rid of excess water through drainage canals or reservoirs, as it is with bringing water to the land.[1] Thus the pattern of irrigation has been, for most periods, one that features "loops," i.e., canals that divide, rejoin, divide, and rejoin repeatedly until the excess water drains into a swamp or one of the rivers. More artificial, designed, brachiating systems have been attempted in certain periods, but have broken down to be replaced by the traditional "looping" system.

The flood in Mesopotamia is tantalizing rather than predictable. The main flow of Euphrates water usually arrives on the plain in April, a bit too late to be of maximum benefit for winter crops. In fact, high water often arrives just in time to ruin young plants and wash away field borders. The flood is also too early for use in summer cropping. In effect, the Mesopotamian farmer, until the modern construction of dams, reservoirs, and the like, faced a yearly

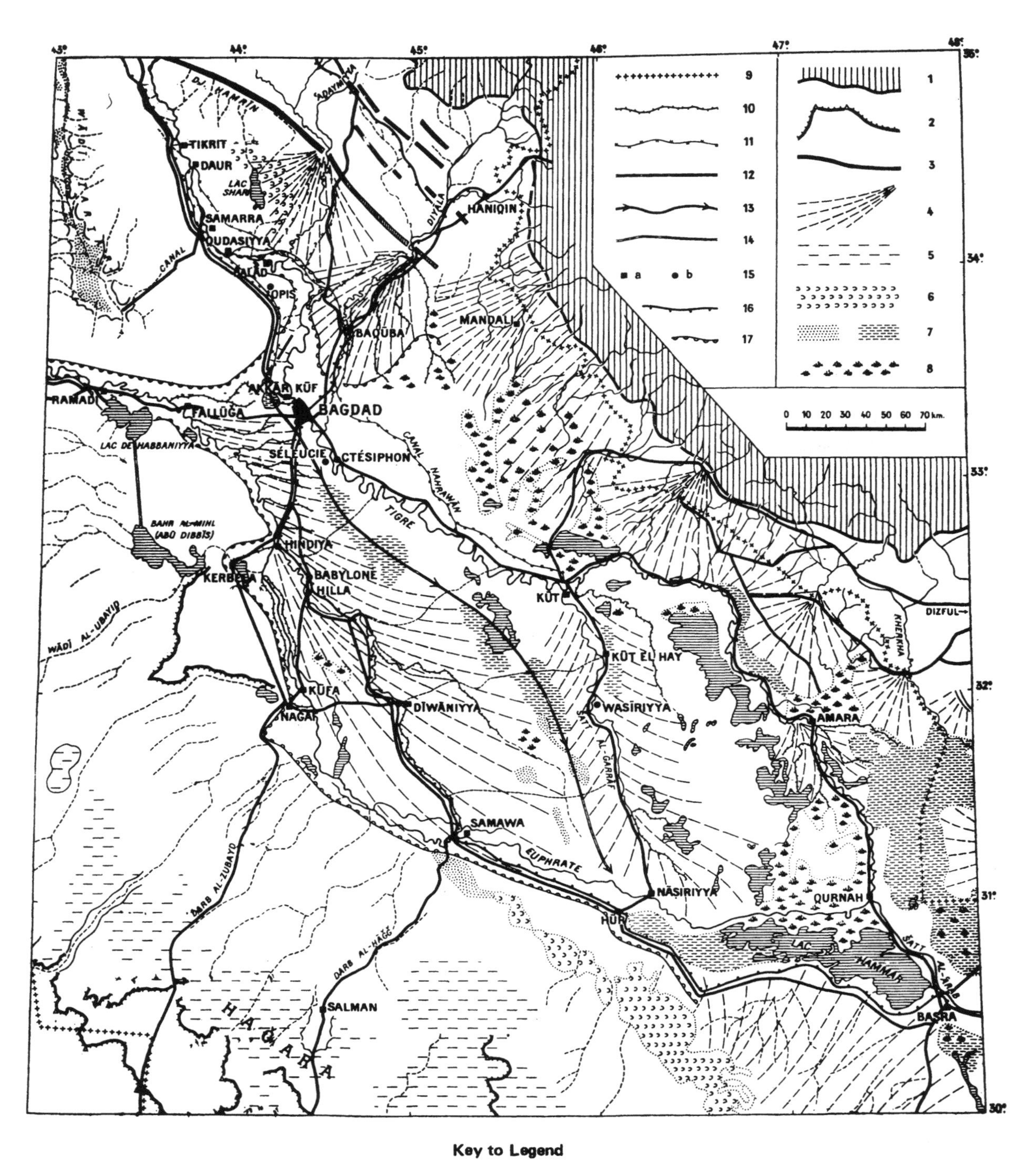


**Key to Legend**

1. Mountains and foothills
2. Declivity
3. Low Hilly Ridge
4. Alluvial Fan
5. Stony Plain, (Passable with difficulty)
6. Dunes
7. Seasonally Flooded Depressions
8. Swamps
9. Political Frontiers
10. Modern Irrigation Canals
11. Abandoned Irrigation Canals
12. Escape Canals for Tigris and Euphrates
13. Series of Depressions used by the Tigris in Flood
14. Roads
15. Towns
    a. Modern
    b. Ancient
16. Railroads
17. Limits of Lower Mesopotamian Plain

Fig. 2.1 Drainage pattern in the Tigris and Euphrates Alluvial Plains. Source: Etienne de Vaumas, 1965, L'écoulement des eaux en Mesopotamie et la provenance des eaux de Tello. *Iraq* 27, pp. 81-99. Plate 21.

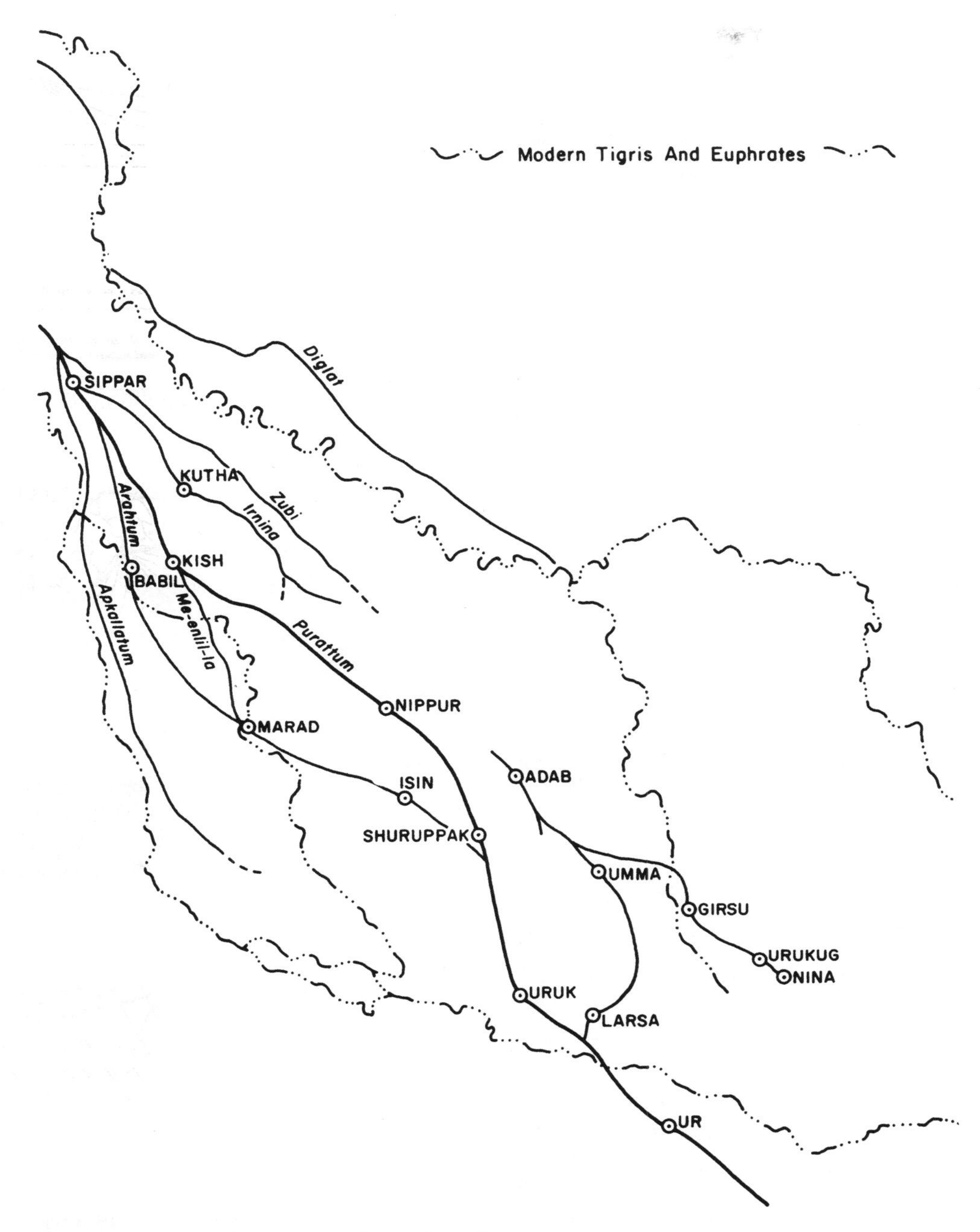


Fig. 2.2 Ancient river systems in southern Mesopotamia.

prospect of previously water-deprived plants being washed away by uncontrollable flooding. Obviously, there was and still is a need to make more water available before the flood by damming streams and canals and raising the water level. The banks of the river and canals must be heightened to lessen the possibility of damage during high-water; however, any damming would be to the disadvantage of people downstream. Likewise, heightening the stream banks at any point could raise the water-level and make the river overflow farther on. In either case, there is likelihood of conflict, deriving from scarcity on the one hand and surplus on the other.

In cutting ditches from natural streams or canals, farmers bring not only water but silt onto their land. Diverting the water slows it down, and silt drops to the bottom. The portion of a channel or ditch nearest the offtake receives more than double the silt that is dropped at the tail (Fernea 1970: 132). However, the silting has a much less adverse effect on the farmers at the head of the canal. Although a reduction in total volume of water carried by the canal will result, irrigators at the upper end continue to take their usual ration of water, while those farther downstream find that they no longer receive full measure. To ensure efficient distribution of water and reduce conflict, the canal must be dredged periodically.[2]

The inequality between farmers near the offtake and those farther along the canal is echoed in soil quality, which is directly related to the process of siltation. The initial slowing of the water causes the largest suspended particles to drop at or near the head of the canal. Progressively finer particles settle to the bottom as the water moves downstream. The coarser grains form a more permeable soil than does the finer material. Thus, the land near the head of the canal tends to drain better. As we will see below, drainage is crucial in dealing with salinization.

Irrigation agriculture in Mesopotamia entails the use of ridges between plots to hold in water. These ridges, as well as canal and ditch banks, act as barriers to natural flow. Thus, basins form where entrapped water from rain, floods, and irrigation may lie for varying lengths of time in different amounts. Ditches or breaks in ridges may allow one small area to dry quickly, while its neighbor remains waterlogged. This process results in a patchwork landscape with variations in soil quality, intimately related to permeability and salinization.

Jacobsen and Adams (1958) have emphasized the role of salinization in the breakdown of ancient civilizations in the alluvium. The salt is derived ultimately from the mountains in which the rivers originate. Through the centuries, salt has been and still is being brought down by the rivers and deposited in the plain. With permeable soils and efficient drainage, the salt would present little problem. It would wash out into the canals after each irrigation and be carried finally to the Gulf. However, the soil, especially at a distance from main streams and canals, tends to be impermeable and the salt accumulates at the water table. If the water table rises, as it does with irrigation, the salt is brought to the surface and its effect is felt on the soil. With modern techniques of drainage, enormous outlays of money, and well-organized controls, such as have been initiated in parts of modern Iraq, the impermeability of the soils can be counteracted and the salts leached out. Although since ancient times (Jacobsen 1958*a*: 64f.) some drainage has been practiced, the elaborate technology requisite to solve the problem has not existed.

Salinization has a direct bearing on the crops planted in the plain. Barley, more salt-resistant than wheat, is the predominant cereal in southern Iraq even though farmers would prefer to plant wheat. This pattern holds true for ancient and Islamic Mesopotamia (Jacobsen 1958*a*, 1958*b*; Jacobsen and Adams 1958). In cuneiform sources, barley most often constituted about 60 percent of the grain yield, and sometimes reached 98 percent. Information on production for ninth century Abbasid Iraq by province shows a gradual shifting from wheat in the north to a barley-dominated region in the south, the line of demarkation being about the latitude of Babylon (Jacobsen 1958*a*: 31ff.). Percentages range from more than 90 percent wheat near Baghdad to more than 90 percent barley near Qurna.

The traditional method of dealing with salinization is a system of alternate-year fallowing. As a result of irrigation, the water table under a field approaching harvest lies at about half a meter below the surface (J. C. Russell, in Jacobsen 1958*a*: 67f.). After harvest, the field turns green with *shok* and *agul* (camel thorn). These wild plants draw moisture from the water table and gradually dry out the subsoil until winter, when they go dormant. In the spring, since the field is not being irrigated, the plants continue to dry out the subsoil to a depth of two meters, thus preventing the water from rising and bringing salt to the surface. Since they are legumes, the plants also replenish the land with nitrogen, and retard wind erosion of the topsoil. In the autumn, when the field is once again to be cultivated, the dryness of the subsoil allows the irrigation water to leach salt from the surface and carry it below, where it is normally "trapped and harmless." However, over a long period, the subsoil trap fills and its top rises toward the surface until the *shok* and *agul* roots lose their efficiency in deep-drying the land. Farmers may dig up and remove the badly-salinized topsoil and work the deeper layers until those too grow unproductive. Then, the land must be totally abandoned for a long period: in the range of fifty to a hundred or more years. Up to that time, however, the fallowing system is "a beautiful procedure for living with salinity. The rural villagers understand it in that they know it works, and they know how to do it and they insist on it" (Russel in Jacobsen 1958*a*: 67f.).

The natural factors outlined thus far – the unreliability of river flood, dependence on flow irrigation and the necessity for canal construction and maintenance, variations in siltation rates and permeability of soils, inequalities in land and the overriding problem of salinization – combine to force on the area an extensive rather than intensive agriculture. By extensive is meant a production system based on areal extension rather than intensification of irrigation or agricultural techniques to give increased production. Enough land has to be available to allow half to remain fallow each year. In such a situation, concentration of resources (good land and water) in the hands of one or a few persons would prove counter-productive since it would yield disproportionate wealth which, if invested in agriculture, would mean intensification of irrigation and agriculture (new canals, summer production, and violation of fallow). When concentration occurs, intensification must result because the mere extension of land, even with an observance of fallowing, will ultimately not bring about enough increased yield, especially if the land-holder wishes to invest in non-agricultural activities. With the fallowing system, it must be remembered that salinization is not negated, only retarded, so that extensions of area tend merely to allow a steady yield, not an increase. Mesopotamia is, then, a land on a treadmill, with priority given to human organizations that allow equality in land tenure and water rights. There is a need for a flexibility that at the same time allows members of a group relative equality and an ability to shift holdings, while giving some sort of structure for the carrying out of irrigation operations requiring organized labor forces. Robert A. Fernea (1970: 54) has argued that in Iraq, "the traditional tribal system of land tenure and use was also well suited to traditional methods of extensive cultivation; indeed the two aspects of agriculture must have evolved together in this region."[3] In order to see that the tribal system is ideally suited to the high-risk conditions on the alluvial plain, it will be necessary to review Fernea's work in detail.

As reconstructed, the tribal organization of the El Shabana cultivators of Daghara, an area north of Diwaniyya in the Hilla district, was originally and is still basically one in which lineages and other kin-based segments formed equal, opposed, and complementary units within a tribe called the Aqra. In this system, which Fernea apparently sees as having once been typical for all southern Iraq (1970: 13), the tribe was the landholding unit, all men being equal, with the *shaykh* (chief) the first among equals. In such a system, there was a check, embodied in self-help, that prevented concentration of land-derived wealth and power. Fernea (1970: 150) says further: "Limited centralized authority within the tribe had been reconciled with the decentralizing tendencies inherent in segmentally structured tribal organizations by the ever-present possibility of revolution, replacing a shaykh from one lineage with a man from another section of the tribe." A shaykh could achieve great power and prestige through his ability to lead in battle and act as a "reservoir of tribal law and an astute judge, . . . enforcing culturally defined and traditional norms. . .[through a] course of arbitration [that] had carefully included the development of support within the tribe for his decisions" (Fernea 1970: 109, 136). Under most conditions he could not pass his power on to a successor. However, in some situations tribal chiefs *could* manipulate the system so as to attain real power and transfer it to another. Part of such a process, Fernea (1970: 105) suggests, may be the development of a dominant lineage which is ". . .the end result of conditions undermining those checks and balances which keep structurally equivalent segments equal. . .[namely] conditions of an agricultural subsistence adjustment where wealth may be accumulated. . . ." We shall return to the possibility of achieving real power without external aid, but for the El Shabana chief, who is also head of the whole Aqra tribe, it is apparent that real power has come from a relationship with centralized government.

It can be argued that prior to effective control of the area by the British in 1922 and an imposition of land-tenure changes through which chiefs became landlords and their tribesmen tenants, the Shabana tribal system was one in which there was a continual shifting of power through raids for land and goods. Land held by a group might not be apportioned equally due to differences in strength (chiefs and their close kin occupied the best land near the canal offtakes), but there was a flexibility built into land-use that made it possible to respond to crisis. The resort to arms, kin ties, and group pressures tended to resolve difficulties. Since settled tribesmen were related to nomadic groups and reconfirmed the ties through marriage, economic exchange, and allowing pasturage on fallow fields, there was also a mechanism, Fernea (1970: 12) notes, for shifting to nomadism if farming became impractical. Animals kept by farmers were, thus, not only a source of income but an ultimate insurance against drought, loss of land, or other crisis.

Besides the recourse to nomadism, the tribal group had the land area requisite for fallowing, a type of land use that allowed shifting of plots to equalize members of the group, and social mechanisms to spread risk and provide for the less prosperous through kin ties or obligations of hospitality. There was, likewise, a capability to coalesce and fragment according to need.

Operations concerned with irrigation, such as cleaning canals and digging new ones, require sizable groups of men for relatively short periods. In usual discussions of centralization of authority, much has been made of the organizational requirements for these tasks. Fernea (1970: 120, 129ff.) shows that regular cleaning of canals is well within the capabilities of tribally organized workmen, and that, in fact, the manpower and time required are remarkably small. Even in constructing canals, indigenous cultivators cope very well, cutting sections piecemeal as new areas are needed for cultivation, swamp areas dry up, or the like. This *ad hoc* kind of construction results in canals that look crooked and unappealing when compared with unit-built, government-planned works. But they suffice.

The ability of a tribal group to fragment into its component parts removes the necessity for sustaining a large number of laborers and for then being forced or tempted to make work for them. It also tends to prevent concentration of power in the hands of any man or group. Under usual tribal conditions, a chief does not *order* units to supply labor or fighting men, but puts out a call and recruits volunteers through force of personality, persuasion, and reference to group solidarity and tradition, often backed by religious sanctions.[4]

Following Fernea's line of reasoning, one sees that the most essential way in which tribal organization is suited to irrigation agriculture in Iraq is its tendency to discourage the concentration of wealth and power. Raiding and forcible occupation of land (which governments see as ruinous) are in fact the mechanisms through which confederations, tribes, segments, and subsegments reestablish a balance. Capitalizing upon such factors as biology, in which a low-ranking family may produce a dozen strong sons or a high-ranking family lose its members through disease, natural disasters such as floods and droughts, and economic failures or successes, groups may reshuffle, move about within an area, or take control for a time. The vulnerability of status forces tribal leaders to turn accrued wealth back to the group in the form of hospitality, help in crisis, and the like. Traditional ideals of equality prevent a chief from investing for personal gain in new irrigation ditches or increased productivity through summer cropping (if the river allows it) or the reduction of fallow land. This dampening of an urge for profit from agriculture is what allows the tribe to continue its extensive, rather than intensive, agricultural practices based on alternate-year fallow.

Tribal cultivation is not the only possible method of producing in the alluvium without violating fallow. If a landlord with a large holding were to allow fallowing (as he could afford to do with that much land) and be satisfied with the yield, without pressing his tenants for greater production; if an external government were to refrain from excessive taxing of the tenants; or if the landlord, pressed for taxes by the government, were not to squeeze his tenants; if the landlord with restraint were to reinvest his return only in his land, canals, and the like; if he were, in short, to be a good landlord (Fernea 1970: 47) and act like a shaykh (chief), it would be possible to carry on productive agriculture without violating fallow and causing increased salinization. However, at the time of his study, Fernea (1970: 12f.) noted that most of the big landowners, who were sometimes absentee shaykhs or speculators residing in Baghdad, opted for profit to invest in the western economy to which Iraq had gained entry, or to spend in luxuries, or to live in Beirut. Other studies (Boserup 1965: 98-99) indicate that this phenomenon is a general one, not restricted to Iraq:

> A strong overlord or central government effectively controls not only that the landlords levy taxes, but also that they pass most of them on to the central authority, which uses this revenue largely for military expense or for urban luxuries. This policy by the central authority is not incompatible with the development of the urban economy, but it is obviously inimical to agricultural investment and tends to result in rural depopulation.

Further, Boserup states that:

When [large-scale irrigation] regions are left in the uncontrolled possession of a landlord class which is either of foreign origin or partner in a precarious alliance with a foreign conqueror, rural investments are in danger of being neglected, because the landlords inevitably go for quick profits and liquid assets. In extreme cases, the result is starvation and depopulation.

The summations quoted from Boserup fit very closely the pre-Revolutionary Iraqi situation. First under the Turks, then under the British, Iraq saw the adoption of landholding "reforms" that encouraged chiefs to become absentee landlords, and resulted in the breakdown of tribal cultivation and the decline of the agricultural base. It may be argued that any form of centralized, urban-oriented government would bring about the same consequences. The foreignness of "overlords" need not be a factor. For a tribal group, which sees all its needs including rights and duties fulfilled in its kin-based system, any external government would be "foreign" and opposed to its interests. A tribal system and centralized government are, in a very real sense, functionally equivalent. It is because colonial powers, such as Turkey and Great Britain, recognized this fact, but assumed that tribal chiefs had real power and dealt with them as if they had, that chiefs were lifted out of context and forced to assume roles that they either did not wish to perform, or carried out to the detriment of their people.

The change in land tenure has been traced by Fernea (1970: 30-37) to the Ottoman Land Code of 1858. Under the reform, anyone occupying land could buy the right to use it from the government. Tribal cultivators, who held the land by tradition and concepts of tribal tenure, refused to pay. The Ottomans generally sold the rights to urban speculators or friendly shaykhs. The latter sometimes sold out to speculators or moved to the city to become absentee landlords. In some areas, such as Daghara, the Turks never had effective control and the tribes were able to carry on much as before.

The crucial changes for the Daghara area can be seen to have resulted from another set of actions by the Ottoman government. In 1870, Midhat Pasha, a vigorous, "progressive" governor, constructed a dam across the Saqlawiyah canal at its source on the Euphrates near Felluja (Fig. 2.1). This work was done in order to prevent the flooding of Baghdad. Although there had been some difficulty with the Euphrates branch that supplied water to Daghara from at least as early as 1850, and although such bad floods as that of 1867 (Fernea 1970: 33) caused alterations in the flow of water, the blocking of the Saqlawiyah forced a much greater burden of water upon the barrage at Hindiyah (Fig. 2.1). This barrage, which was the critical divisor of water, allowing a flow to go into the Hilla channel and thence into the Daghara canal, gave way and, although repaired, continued to malfunction. The result was that for the last part of the nineteenth century the Daghara area suffered crises of water shortage. A disastrous and final breaking of the barrage occurred in 1903. The effects of this collapse were vividly recorded by H. W. Cadoux (1906), who rode in a stage coach down the middle of the dry Hilla bed. He tells of whole sections of the countryside being deserted, the former residents having relocated on the Hindiyah channel. Only a few forts, dependent on wells sunk in the middle of the canal beds, were still occupied. All vegetation except palm trees had withered.

Restoration of the dam in 1914 partially alleviated the problem, but the Daghara area did not fully recover until the British reexcavated major canals and put a new regulator on the Daghara canal in the 1920s.

The massive disruptions of people, the fighting, and the relocations on other channels as a result of the river change have not been detailed in any report.[5] For the Daghara area, Fernea (1970: 34f.) gives an outline of events which can be supplemented from other sources. For some time prior to the catastrophe, the Turks had a policy of encouraging smaller tribal groups to take land claimed by large tribes. Thus, when the powerful Khazail tribe moved from the affected Daghara area to other holding on still functioning channels, Shaᶜlan, a minor lineage head of the Shabana segment of the Aqra tribe, appropriated Khazail land along the canal. When water was restored to the area and the former holders tried to move back, they found the Aqra in possession, backed by the Turks, with Shaᶜlan and later his son Atiya as tribal chief.[6]

Under previous conditions, Shaᶜlan's lineage would have held sway for some time, but its hold would have been precarious and another leader would eventually have taken the shaykhship. However, historical events occurred which shifted positions and roles and began the process by which the Shabana became, as Fernea found them in the 1950s, tribal cultivators distorted in function by the changing of the chief's role from leader to landlord.

The advent of real control of the countryside by the British, accompanied by a policy of deliberate backing of landlords, was the first element in the change. Chiefs were regarded as having real power at the same time they were named landlords over tribesmen. Control of the country, through control of the shaykhs, was thought to be productive of peace as well as agricultural yield. The second important element in the change was the reactions of the chiefs to government control. In 1918, an official report on the Aqra tribe said there was no paramount chief (usually meaning the head of a dominant lineage), but that power was being contested by Saᶜdūn al Rasān, head of the Hamad segment (which still holds land across the Daghara canal from the Shabana), and Shaᶜlan. In another report, however, Shaᶜlan is said to have had his tribe well in hand, to be popular, and to have ". . . a high reputation for straightforwardness. He has been consistently well-behaved and helpful since the occupation" (Arab Bureau 1918: 199; see Fernea 1970: 138). All other shaykhs in the regions were reported in a bad light. I would suggest that Shaᶜlan, having effected a hold on the land under the Turks and wanting to consolidate it, began to collaborate with the British in the interests of his section and his lineage. His title as chief of the Aqra seems to have been made secure by this policy. It is not clear exactly when Atiya took over the shaykhship from Shaᶜlan, but that Atiya continued to function as a tribal leader is attested by his leading revolts as late as the 1930s. Thereafter, the British replaced him with his son Mujid, who was more tractable.

In the 1950s, Shaykh Mujid, as a medium-sized landholder, was still on his land and performing some tribal functions. Because he was doing so, members of his tribe remained as his tenants although clearly they would have been more prosperous as sharecroppers for a larger landowner (Fernea 1970: 50f.). The tribal ideal was strong enough to offset a certain degree of economic deprivation. Seasonal labor in the cities, or other work, was used to supplement farm income and allow men to remain on Shabana land.

The reason why sharecropping on a large holding had become the most advantageous alternative is tied to fallowing; only a large landowner had sufficient acreage to shift tenants around in order to reduce losses due to poor land, and allow full fallowing. Tenants on medium-sized holdings (e.g., the Shabana) and small farm owners were forced to violate fallow to meet rents, taxes, and debts. Since Iraqi farmers realized a good part of their income and nourishment from their animals, any reduction in fallow, which is pasturage, would entail a decrease in livestock. The importance of animals to the cultivators is indicated by the net income derived from them: 66.8 percent for tenants on large holdings, 56 percent for tenants on medium-sized holdings, less than 50 percent for owners or tenants on orchards, and 42 percent for small farm owners (Poyck 1962: 84).

The change in tenure practices and the gradual disintegration of the agricultural and tribal system are focused in the changed role of Shaykh Mujid. The government, under the British and later under the Kingdom of Iraq, not only made him a landlord and gave him a disproportionate share in the wealth, but also guaranteed him his position. In that role (Fernea 1970: 109, 136) he

> ...as an arm of the administration was placed in the position of enforcing regulations which often totally lacked any basis in tradition. Being required to carry out jural duties without having opportunities to demonstrate those qualities which traditionally permitted him to assume jural responsibilities in the first place, gradually undermined rather than strengthened the shaykh's authority among the tribesmen.

He had power but not authority. The result was that disputes over water rights, rather than being settled finally by the shaykh, were being taken to the government irrigation engineer. The imposition of civil law, the arrival of various administrators, and the opening of a school, all contributed to remove from the shaykh functions he would have carried out. In this curious position, the shaykh began to act increasingly in his own family's interests. The initial tendency of other tribesmen seems to have been to do the same. When the British began to register land within the tribal area, it was usual for men to register in their own names (Fernea 1970: 98-99). However, in the following generation, men registered jointly as brothers, and in the third generation registered together as brothers and patri-cousins. The reassertion of kin ties must be seen as a realization that division into smaller and smaller plots reduced the possibility of fallow, raised the probability of salinization, and was in other ways counterproductive. The recombining of plots was an obvious attempt to repair the damage and return to an ecologically sound land use.

That small holdings did damage the land is supported by figures showing that small landowners had a disproportionate amount of salted or otherwise useless land (Fernea 1970: 45ff. 148f.). However, the general rise in salinity may be traced in part to another cause: the opening of a major new canal by the central government and the maintenance of high water through the year by construction of upstream storage basins and dams.

Thus, directly through engineering that promoted increased waterlogging and salinity, the central government acted to undermine agricultural productivity. However, the effect of the government's indirect intervention, through pressure on the tribal system and change in tenure laws, was probably more deleterious. Merely by supporting and keeping one family in a position of power; by changing a chief to a landlord; by concentrating wealth while inducing individuals to take up small, fixed plots; by imposing yearly taxes and encouraging rents and debts, the central authority brought about widespread violation of fallow. Eventual selling out by small holders to large landowners did not lead to a reversal of agricultural decline because debt-ridden farmers often did not stay on the land as sharecroppers, but became nomads or fled to the cities. Fernea (1970: 52) describes the situation: "In spite of the cooperation of the police with the landowners, it became comparatively easy for sharecroppers to run away to Baghdad. So great was the need for sharecroppers that estate owners accepted 'runaways' from other areas without asking too many questions." The wealth and strength of the central government, reflected in control of the countryside and irrigation schemes, had opened alternative employment. The result was a labor shortage in the countryside. Tenants brought in from other areas tended to be less effective than tribesmen had been. For their part, Fernea (1970: 42) states, "the tribesmen [were] anything but committed farmers. . . .Many would rather do almost anything else. . .through discouragement from struggling in a trying climate with poor land, high rents, and scarce water." Clearly, the ruination of the countryside, which Boserup sees as a general consequence of landlordism in irrigated areas, was in progress when Fernea was carrying out his study.

It may be argued that for Iraq, government's best course has always been merely to see that water reached tribal cultivators and let them handle it from there. That this method, historically, has not been followed is evident. As early as 2400 B.C., a drastic increase in salinity in southern Iraq seems to have been linked to the cutting of a new irrigation canal by Entemena of Girsu (Jacobsen and Adams 1958: 2). By 2100 B.C., the salinization had spread ". . .westwards towards the Euphrates through the heart-land of the South" (Jacobsen 1958*a*: 2). The area never recovered and hegemony passed to Babylon in the eighteenth century. The northern part of the plain experienced a salinization crisis between 1300 and 900 B.C. (Jacobsen and Adams 1958: 2). As power moved north, so did the salt problem.

In the Diyala region, the Islamic construction of the Qatul canal, which brought an overabundance of water into the area and made summer crops possible, encouraged over-irrigation and the subsequent abandonment of wide areas. Jacobsen (1958: 8-9) notes that in areas not watered by the Qatul, cultivation has continued to the present day.

Probably the most destructive engineering feat in ancient Mesopotamia was the cutting of great transverse canals that linked the Euphrates and the Tigris (Le Strange 1895: 71ff.). These canals, begun shortly after the Muslim conquest, were wide enough to permit large sailing ships to pass through them. The amount of water in these canals, affecting the area from Baghdad to Babylon, would have raised the water table dramatically and caused salinization at a tremendous rate. The effect of these canals may be seen in the fact that although in the ninth century A.D. this area was producing from 50 to 98 percent wheat, by about 1250 the southernmost canal, the Nil, was out of use, and by about 1500 almost the entire area was deserted. In the 1950s, the region became the site of the Mussayib Project, a special drainage scheme designed to deal with salinization.

It will be objected that it was not salt that led to the abandonment of these areas, but rather the weakening of central control. It is true that a slackening of administration would have made silting, which on the vast, planned projects would have been tremendous, a difficult problem to handle. However, I would argue that the breakdown in administration was merely a late phrase of a much larger cyclical process that has seen the rise and fall of kingdoms since the fourth millennium.

The cycle, in my reconstruction, was first triggered by a natural disaster occurring at the end of the Uruk Period (c. 3200 B.C.). This event, which as I have argued elsewhere (Gibson 1973) brought about the earliest development of true urbanism, was a change in the Euphrates that left dry a large section of the plain (see Adams 1972: 740) as the water shifted into other channels from a branch that had run east of the one marked "Zubi" on Figure 2.2. Although in the town of Uruk there had been moves toward centralization of authority and other features usually connected with civilization, they did not exist on a *general, regional* scale until the inhabitants of the affected channel were forced to crowd in upon occupants of the other courses. The sudden increase in population density was a crisis that forced into existence new social relationships and organizations through competition for land and water. In this view, although natural population increase would have eventually brought about the development of civilization, the river shift acted as a catalyst. The key element in the change is proposed to be, as with Adams (1955, 1966), the growth of social stratification.

The mechanism behind the evolution of stratification involved the change in role of a traditional leader, through warfare, into a supratribal king holding real power (as with Adams 1955, 1966). The shifting of population from the abandoned stream to already occupied lines would, in my reconstruction, have entailed the exaggerated reliance by both newcomers and old occupants on their respective traditional leaders. The steps by which such a leader, by clever manipulation of the tribal system and the crisis situation, could rise to real power may, perhaps, be elucidated through studies of men who have transformed tribal polities into kingdoms. Such a case is that, given by H. Rosenfeld (1965), of Ibn Rashid, who formed a kingdom in the tribal setting of nineteenth century Arabia. Ibn Rashid, the head of a dominant lineage in a vigorous tribe, first captured the town of Hail. Then, by a deliberate policy of rewarding non-kin, such as slaves, townsmen, and members of lineages not closely connected to him, he began to form a striking force and a retinue of men loyal only to himself, rather than to a kin group. At the same time, by blocking his close kinsmen from positions of power, he effectively reduced their influence and prevented the consolidation of opponents around a rival. He had the option of using or not using his kinsmen. The actions of cutting his kin ties, while forming alliances across lineage lines and outside lineage contexts, as well as creating an independent striking force to act as sanction for his decisions, made him a king rather than a chief. It is interesting to note that in Fernea's study (1970: 106ff.), once having taken the shaykhship, Sha<sup>c</sup>lan and later Atiya followed a similar course, broadening their base by marrying off sons to daughters of other lineage heads, especially those who had been most injured by the takeover.

In ancient Mesopotamia, once polities that required taxes or some other levy to support warfare, trade, and the like had come into existence, the agriculturist would have been pressed to produce more. Irrigation schemes, new canals, and administration of the entire system would have brought about changes in the landholding and social systems. Landlords, or temple or palace establishments as landholders, would gradually have replaced smallholders. The subsequent flight of sharecroppers to the cities or into nomadism, importation of outsiders to do the farming, further pressure from the government to produce food for the growing population in the cities, a gradual deterioration of agricultural land, reduction of investment in irrigation because of low returns, and breakdown of canal systems would follow. The state might be able to exist for long periods after the decline of the agricultural base, if warfare or trade could bring in food or slave labor (but slave labor is inefficient for irrigation agriculture). However, eventually, because of salinization, the state must fall or be toppled by stronger neighbors. With that demise, tribalism and agriculture would revive.

In the history of the Mesopotamian plain, the damage done to agriculture, and eventually the plain itself, by the central administration would have varied according to how strong the government was, how intimately it could make its will felt throughout the countryside, and how deeply it was involved in irrigation. Variations in the degree to which cultivators were allowed to administer their areas would have affected the speed with which salinization took place. It would not be surprising to find that in almost any period there was a variety of landholding, e.g., palace, temple, big landowner, small landowner, tribal holdings, much as is described by Poyck and Fernea. However, the percentages of types of landholders would vary with the area or the phase of deterioration reflected in the records. The mere fact that irrigation records were being kept is an indication that the process was under way.

Usual considerations of cuneiform records either entail the assumption of too formalized, bureaucratic, legalistic structures (everyone is an "official" or is assumed to have only one role), or, more surprisingly, treat Mesopotamian culture as one of free enterprise, individual initiative and little active bureaucracy at all.[7] Kinship systems are usually allowed for nomads, but settled people are seen as unaffected by kin ties, lineage obligations, and other such ties. An example of a study in which officials abound is one by Walters (1970) concerning irrigation around Larsa in the nineteenth century B.C. One wonders what would have been made of the several "supervisors," "overseers," "canal contractors," and the like had

Walters read Fernea. Several situations in Walters' material seem very familiar. State agents are sent into an area to survey and build canals with local labor. These local men, who may, I think, safely be called tribesmen or at least kin-organized groups, are to be recruited by local men who are, in my opinion, tribal figures. One of these, Lu-igisa, a man of wealth, writes directly to higher officials concerning grievances and seems to be in somewhat the same position as Fernea's shaykh. Namely, he is a local leader whose role is in a process of change, partly coopted by the government and even given a title, but still carrying on some traditional functions. Some texts in the archive resemble accounts of cases settled by Fernea's irrigation engineer (1970: 166-69).

I would suggest that most ancient records of agriculture and irrigation in Mesopotamia would fit into the context of the process described by Fernea. With a sizable archive, covering a long span of time, one should be able to discern a rising degree of centralized administration and government-planned canals, concurrent with an increase in local social disintegration. As time goes on, the amount of land in fallow should decrease. Initial rises in production, a result of violation of fallow, should be followed quickly by a lowering of yields. Contemporaneously, the proportion of smallholders should become fewer and largeholders greater. More freemen should become share croppers or "serfs," and the percentage of runaways should increase with time. I would further suggest that eventually, in strongly centralized kingdoms, as a result of the landlord's inability to control labor, landholding would have come increasingly into the hands of the state or a combination of state and temple. I would also propose that the longer the process was in operation, the greater would be the number of officials concerned with irrigation and rural administration.[8] This proliferation of bureaucrats would come about through attempts to remedy the damage to agriculture through new engineering schemes and reforms in administration. Concurrently, later phases should be marked by a cheapening of titles, changes in the functions of officials, and a confusion in lines of command.

Whether evidence to support such notions can be gathered from texts is subject to test. It cannot be denied, however, that Fernea (1970: 153) has given us a very valuable framework in which to view Mesopotamian civilization: "the congruence of fit between tribal methods of cultivation and land tenure and the nature of land, water, and climate" is clear. Whether one can go beyond Fernea and assert the validity of the cycle implied by his work must be investigated in the written sources. I would insist that a total return to a tribal system after each decline in centralization would not have occurred. However, enough of the land would have returned to traditional methods of cultivation to ensure the recovery of the agricultural base. It is evident that, historically, Mesopotamian civilization did not go back to zero after each kingdom's fall. Rather, it can be shown, if through nothing more sophisticated than a measurement of territory held, that a decline of a state was followed by the establishment of a larger, more powerful, more complex one. This replacement at a higher level may be explained by the fact that when disaster struck, destruction was not total; some parts of the old system continued to function. When a new elite set out to rebuild the system, the reconstruction was done on a planned basis, not piecemeal. The emergency operations were more planned and more artificial than the naturally developing ones. In a normal situation, until something fell apart completely, it was kept going by small repairs, much as was the case with the Hindiyah Barrage. In the collapse and reconstruction of a civilization, a state, an irrigation system, or any other large corporate entity, because of the mixture of the old and the new, there emerged a new set of subordinate and superordinate relationships. New social structures were imposed and interwoven with old ones. This complex layering of relationships and organizations is at the base of what is called "the evolution and elaboration of social stratification." The ecological approach given by Fernea, combined with the time depth afforded by cuneiform and other records in Mesopotamia, should make possible richer syntheses of that evolution and the processes of civilization.

## NOTES

[1]On the antiquity of drainage schemes, and the recognition of need for them, see Jacobsen (1958*a*: 64ff.).

[2]Of great importance as a contribution to blockage of canals is the dust carried by the numerous sandstorms in Iraq each summer.

[3]That the fallow system is ancient, has been shown by Jacobsen (1958*a*: 65f.). In harvest records for Girsu, dating from about 2500 B.C., specifically named fields appear every other year. Likewise, there is an indication, although regrettably negative evidence, for an observance of the ban against summer cropping. This lack of summer cropping may be explained partially by the fact that in order to produce enough water, extensive reservoir and damming systems would have to have been built, and at that time, such technology was lacking. It should be noted, however, that even in the 1950s, a period when a vigorous central government was encouraging increased production through summer cropping, summer production made up less than 4 percent of agricultural production in the Diyala Region (Adams 1965: 13) and for the Daghara-Hilla area only 15 percent (Poyck 1962: 84).

[4]Fernea (1970: 130) describes such a call for canal work in which the chief's son or another representative went around to various hamlets accompanied by a respected holy man. Peters (1897: 240) describes a tribal dispute in an area just east of Daghara in which two sections of a tribe went to war. He notes that after an initial incident, one group "...returned in force, accompanied by an imam – sure sign of mischief – for the purpose of attacking [the other group]." The imam, a holy man, was obviously carrying out a similar role in a warlike situation.

[5]A slight reference is made to it by Gertrude Bell (1918: 128). "The half-settled tribesmen [of the Hilla branch] moved across to the Hindiyah, with an accompanying social dislocation which it is difficult to picture."

[6]From reports of British administrators (e.g., Great Britain, Arab Bureau 1919: 26), it is clear that Sha$^{c}$lan and Atiya had taken control of the Aqra tribe from other segments, specifically the El Ghrush and the El Shawahin. In Fernea's study (1970: 206), these two groups received the least water and were in other ways at a disadvantage. Since Fernea does not mention that these two groups were formerly shaykhs' lineages, it may be that some time had passed since they had been so. Fernea says that Atiya took the shaykhship from the Elbu Ubayd, who were in the 1950s in a slightly higher status than the El Ghrush and the El Shawahin. It is important to note, however, that in trying to heal wounds and cement alliances, Atiya and his sons contracted marriages with all those "...who suffered considerable loss of land and position [except for] ...El Ghrush and El Shawahin..." (Fernea 1970: 106). Regardless of the sequence of shaykhship, the information that can be gleaned about the takeover is a good illustration of the equalizing, balance-redressing feature of a tribal system.

[7]I owe to Mr. Norman Yoffee the notion that some scholarly assessments of ancient Mesopotamia have a basic assumption of non-directed, non-structured economy. Administration, as an organized structure, has almost never been dealt with. Mr. Yoffee, in his forthcoming doctoral dissertation, will outline the bureaucratic structure in the Old Babylonian Period.

[8]Mr. Yoffee informed me that his research on Old Babylonian administration shows precisely this phenomenon, but on a wider scale than merely agriculture and irrigation. There was an increase in the number of officials as the dynasty grew weaker.

## REFERENCES


ADAMS, Robert McC.

1955 Developmental Stages in Ancient Mesopotamia. In Julian H. Steward, editor, *Irrigation Civilizations: A Comparative Study.* Washington, D. C.: Pan American Union, pp. 6-18.

1958 Survey of Ancient Watercourses and Settlements in Central Iraq. *Sumer* 14:101-104.

1965 *Land Behind Baghdad: A History of Settlement on the Diyala Plains.* Chicago: Univ. of Chicago Press.

1966 *The Evolution of Urban Society: Early Mesopotamia and Prehistoric Mexico.* Chicago: Aldine.

1972 Patterns of Urbanization in Early Southern Mesopotamia. In *Man, Settlement and Urbanism*, Peter J. Ucko, Ruth Tringham and G. W. Dimbleby, editors. London: Duckworth.

[BELL, Gertrude]

1918 *The Arab of Mesopotamia.* Basrah: Superintendent, Government Press.

BOSERUP, Ester

1965 *The Conditions of Agricultural Growth.* Chicago: Aldine.

CADOUX, H. W.

1906 Recent Changes in the Course of the Lower Euphrates. *The Geographical Journal* 28:266-76.

FERNEA, Robert A.

1970 *Shaykh and Effendi: Changing Patterns of Authority Among the El Shabana of Southern Iraq.* Cambridge, Mass.: Harvard Univ. Press.

GIBSON, McGuire

1973 Population Shift and the Rise of Mesopotamian Civilization. In Colin Renfrew, editor, *The Explanation of Culture Change: Models in Prehistory*. London: Duckworth, 447-63.

Great Britain Arab Bureau

1918 *Administration Report of Diwaniyah District, 1918.* Baghdad.

1919 *Arab Tribes of the Baghdad Wilayat, 1918.* Calcutta: Superintendent, Government Printing.

JACOBSEN, Thorkild

1958*a* *Salinity and Irrigation Agriculture in Antiquity. Diyala Basin Archaeological Project, Report on Essential Results June 1, 1957 to June 1, 1958* (Mimeographed). Baghdad.

1958*b* Summary of a Report by the Diyala Basin Archaeological Project, June 1, 1957 to June 1, 1958. *Sumer* 14, pp. 79-89.

JACOBSEN, Thorkild and R. McC. Adams

1958 Salt and Silt in Ancient Mesopotamian Agriculture. *Science* 128:1251-58.

LE STRANGE, Guy

1895 Description of Mesopotamia and Baghdad written about the year 900 A.D. by Ibn Serapion. *Journal of the Royal Asiatic Society,* 1895, pp. 1-76, 255-315.

PETERS, John P.

1897 *Nippur, or Explorations and Adventures on the Euphrates.* New York: Putnam.

POYCK, A. P. G.

1962 *Farm Studies in Iraq.* Mededelingen van de Landbouwhogeschool te Wageningen, Nederland, Vol. 62/1. Wagneingen, the Netherlands.

ROSENFELD, Henry

1965 The Social Composition of the Military in the Process of State Formation in the Arabian Desert. *Journal of the Royal Anthropological Institute of Great Britain and Ireland* 91:75-86, 174-93.

WALTERS, Stanley D.

1970 *Water for Larsa.* New Haven, Conn.: Yale Univ. Press.

YOFFEE, Norman

n.d. *The Economic Role of the Crown in the Old Babylonian Period.* Ph. D. Dissertation, Yale Univ. In Preparation.

CHAPTER 3

# SASSANIAN AND EARLY ISLAMIC WATER-CONTROL AND IRRIGATION SYSTEMS ON THE DEH LURAN PLAIN, IRAN

James A. Neely

*Department of Anthropology, University of Texas at Austin*

The data presented in this paper were derived as part of a comprehensive, ongoing archaeological program conducted on the Deh Luran Plain of southwestern Iran. Sponsored by Rice University and funded by the National Science Foundation, this program has been under the general direction of Frank Hole. The majority of the data was collected during my 1968-69 reconnaissance of the Plain (Neely 1969), although some information comes from our 1961 and 1963-64 field work as well (Hole, Flannery, and Neely 1969).

This paper will relate briefly the results of the initial analysis of a portion of the data recovered. In this regard I am endebted to John Hansman of the University of London for his aid in ceramic identifications. While the basic information and associations herein described are accurate to the best of my knowledge, the reader is forewarned that the temporal placement of some of the sites involved is as yet uncertain. I feel the trends brought to light in the initial analyses will stand as presented. However, as I am presently seeking the aid of certain specialists to assess the results of my initial analyses of materials with which I am unfamiliar, it is possible that changes will have to be made in such categories as mean site size and population estimates.

It is partially the result of my own enthusiasm to make known the information at hand, as well as the earnest persuasion of several colleagues, that I am presenting this paper at this time rather than waiting for the final manuscript covering the entire survey to be completed for publication. I particularly wish to thank Robert McC. Adams and McGuire Gibson for their encouragement in this regard, but they most certainly should not be held accountable for my analyses and any mistakes purveyed herein.

## ENVIRONMENTAL SETTING

The Deh Luran Plain (Fig. 3.1) is located in southwestern Iran, near the border with Iraq, some 300 kilometers northwest of the Persian Gulf and 550 kilometers southwest of Tehran. Note that this is about 200 kilometers southeast of the Diyala region of Iraq and 125 kilometers northwest of the Upper Khuzistan Plain – these are areas to which I will refer later in the paper in relation to the findings of the intensive archaeological surveys conducted by Robert Adams (1962; 1965).

The Deh Luran Plain lies within the semi-arid Assyrian Steppe (Hatt 1959) biotic province at an elevation of about 150 to 300 meters above sea level. The annual precipitation of 250 to 350 millimeters (9.5 to 13.5 inches) is by no means equally distributed throughout the year. The summer months are dry and hot, with high mean temperatures of over 120° Fahrenheit quite common. In the winter, when the vast majority of the precipitation occurs, the alluvial plain is transformed in places into meadows of various grasses and wild flowers.

However, the Plain is not a uniform or homogeneous environmental zone. Through personal observation and in consultation with the project's geographers, botanist, and zoologist at least four microenvironmental zones may be defined.[1] These zones (Fig. 3.2) are based on present-day situations (Hole, Flannery, and Neely 1969; Kirkby and Kirkby 1969), as well as data derived from studies pertaining to the early periods of occupation of the Plain (Hole, Flannery, and Neely 1969). The four zones are:

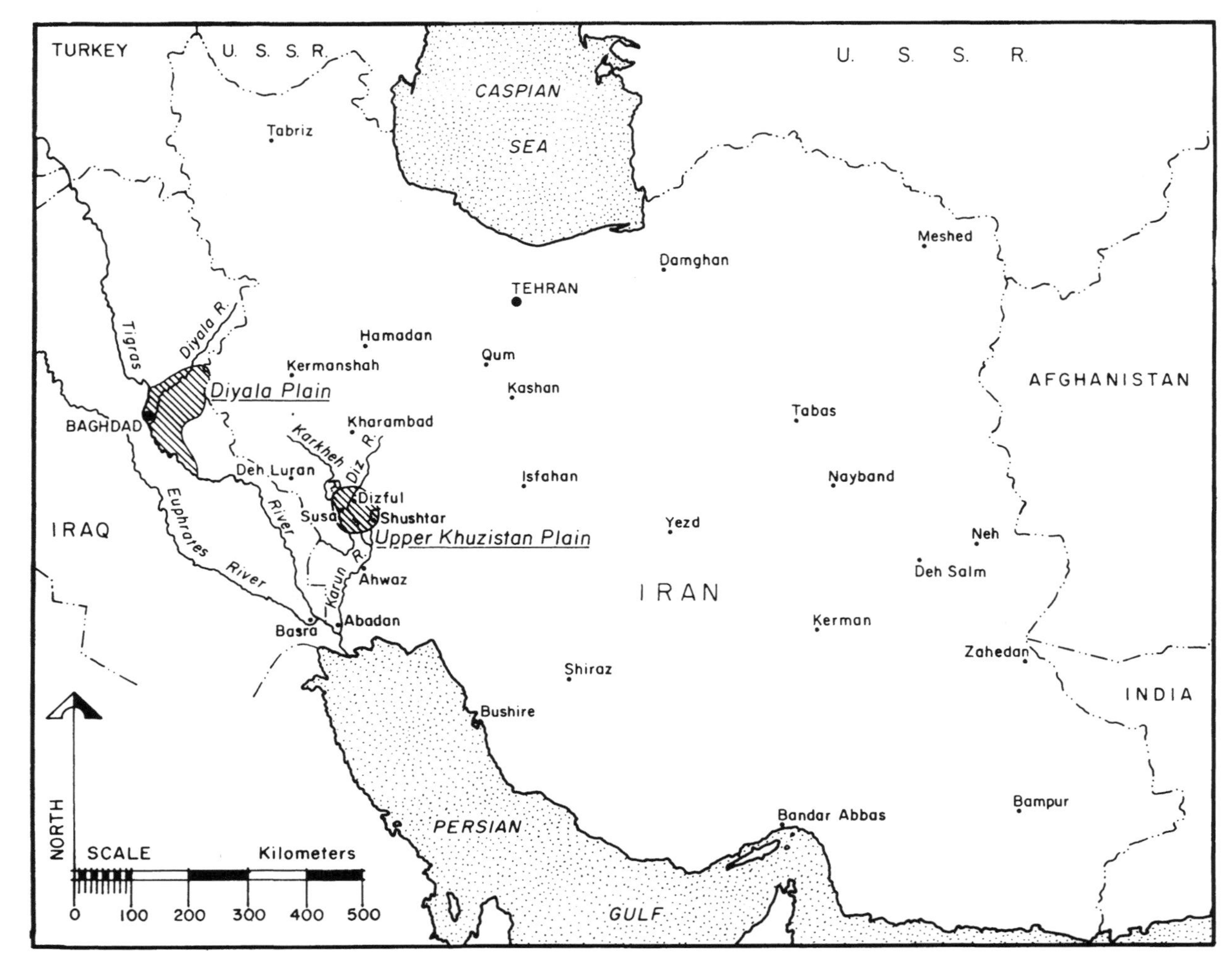


Fig. 3.1 Deh Luran in relation to the Diyala Plain and the Upper Khuzistan Plain. The villages of Nayband and Deh Salm (Spooner, this volume) are located in east-central Iran.

1) alluvial plain zone
2) shallow, salty marsh zone
3) riverine zone
4) rocky piedmont zone

1) The *alluvial plain* forms the largest of these microenvironments. It is essentially a flat plain with small amounts of scattered, low vegetation. Natural topographic features consist of "sink holes" or depressions and the erosional cuts of the two main perennial river systems, a few smaller spring-fed water courses, and numerous intermittent washes. Man-made features consist of canals and qanat systems, and nearly all forms rising above the level of the plain's surface.

2) The *shallow, salty marsh* extends as an L-shaped zone divided into two segments. Lying predominantly in the west-central portion of the plain, the marshy area extends southeast until it takes a more easterly trend. The second, smaller segment, which forms the easternmost extreme of the L-shaped zone, lies across the Dawairij River at a distance of some seven kilometers. The zone ranges from about one to three kilometers in width and is approximately thirty kilometers in total length.

Characterized by saline soils and sparse low vegetation of highly salt-resistant shrubs, the zone is dissected by numerous erosional cuts. While marshy during the winter rainy season, this microenvironment is essentially dry and salt-encrusted during the summer.

This marsh area apparently served, as it does today to a limited extent, as a drainage area for excess irrigation waters. The sparseness of habitation within

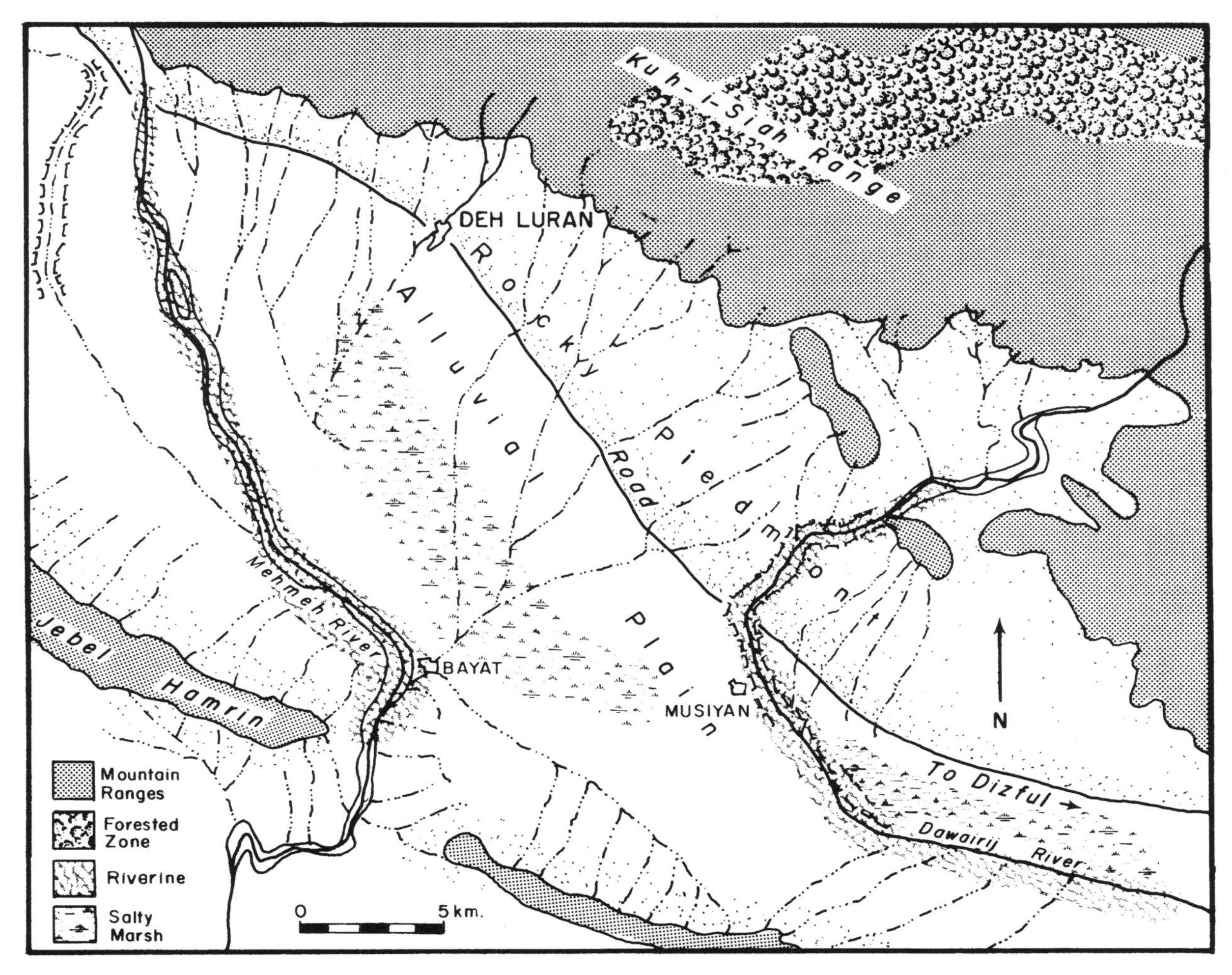


Fig. 3.2 The Deh Luran Plain, showing the approximate boundaries of the four microenvironmental zones defined in the text.

and immediately surrounding this area reflects its lack of usefulness in the past.

3) The *riverine microenvironment* consists of the channels of the Mehmeh and Dawairij rivers and their flood plains. The rivers, especially the Mehmeh, carry brackish waters charged with gypsum and other salts from geological strata cut by their channels in the foothills and mountains north of Deh Luran.

The terraced flood plains, lying between the main river channel and the surface of the alluvial plain, are most extensive in the southern portion of the survey area. Here the flood plains are vegetated rather densely with grasses and low forests of Tamarisk (*Tamarix*), Licorice (*Glycyrrhiza*), and Poplar (*Populus*).

4) The *rocky piedmont* lies north of the improved road passing southeast-to-northwest through the Deh Luran Plain from Dizful, the closest urban center some 125 kilometers to the east-southeast.

The piedmont rises in a fairly steep slope to the foothills of the Kuh-i-Siah Range of the Zagros mountains. This microenvironment is quite rocky and highly dissected by erosional channels.

A number of small springs effloresce near and at the juncture of the piedmont with the foothills of the Kuh-i-Siah. Compared with the alluvial plain, the piedmont is characterized by a substantial increase of perennial grasses as well as small trees and shrubs.

## THEORETICAL ORIENTATION

A number of problem orientations in the form of hypotheses were formulated for testing by means of the then proposed survey of the Deh Luran Plain. Many of these hypotheses seem simplistic and almost too obvious to be of any use. However, as the methodological and theoretical frameworks of Middle Eastern archaeology have lagged behind those used in other parts of the world, we are faced with a relatively large number of implicit assumptions at all levels of investigation. The formulation of hypotheses was an attempt to put some of these assumptions and other ideas to the test by means of a more scientifically rigorous procedure.

In considering the probable presence of a Sassanian and Early Islamic occupation on the Deh Luran Plain, a hypothesis was formulated on the basis of Adams' (1962; 1965) settlement pattern data from the not-too-distant Diyala region to the west in Iraq, and the Upper Khuzistan Plain to the east.*

In brief, this hypothesis stated that, considering the available natural resources of the Deh Luran Plain, should the Sassanian Period witness a dramatic increase in population density with a concomitant change to more numerous sites and a dispersed site and settlement pattern, we should expect a corresponding increase in the efficiency of water-control and irrigation techniques. In this case, such modifications would be expected not to be merely a reflection of the changes in settlement patterns, but a means by which an increasing population utilizing a new, but as yet not well-known or understood sociopolitical system could adapt itself to the Deh Luran Plain.

## PRESENTATION OF DATA

Over three hundred sites, including water-control and irrigation systems as well as habitations, were recorded by the survey. While the entire 1000 square kilometer-area of the Plain was not surveyed, the use of what I call "zone" and "band" sampling techniques resulted in the location of an estimated 80 percent of the visible sites and features in each of the microenvironmental zones defined (Neely 1969: 9-11). The temporal placement of sites is based primarily on the comparative study of the ceramics recovered during the survey. This study was greatly facilitated by Robert McC. Adams' (1970) recent work at Abu Sarifa and through the good offices of John Hansman.†

The sites and features of the Achaemenian, Parthian, Sassanian, and Islamic Periods were conspicuous on the surface. Not only could they be located with relative ease, but with a minimum of trowling it was often possible to determine and accurately map the various components of the sites (Fig. 3.3), and even the internal features of many of the houses (Fig. 3.4).

This ease of location and mapping is probably the result of a number of factors. The relatively recent date of these features (about 300 B.C. to A.D. 1250), the nature and location of their construction, and the fact that the project geographers have determined

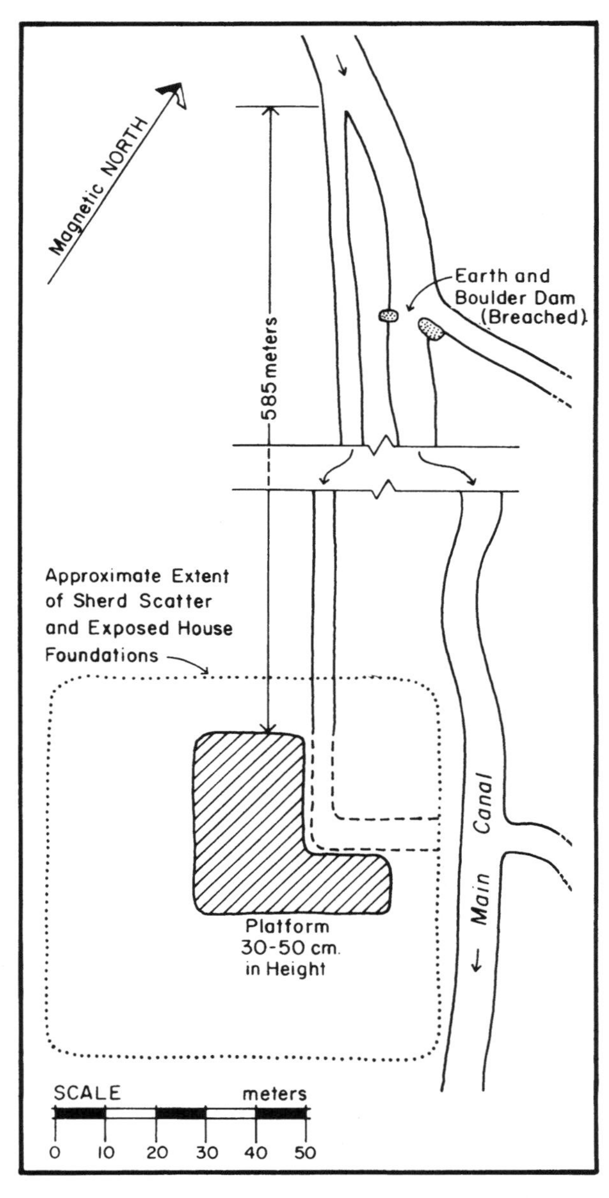


Fig. 3.3 Site DL-17, a homestead or a hamlet on the Alluvial Plain, showing the associated canals that provide domestic water supplies and water for irrigation.

*Robert McC. Adams 1968: personal communication.

†John Hansman 1968-69: personal communication.

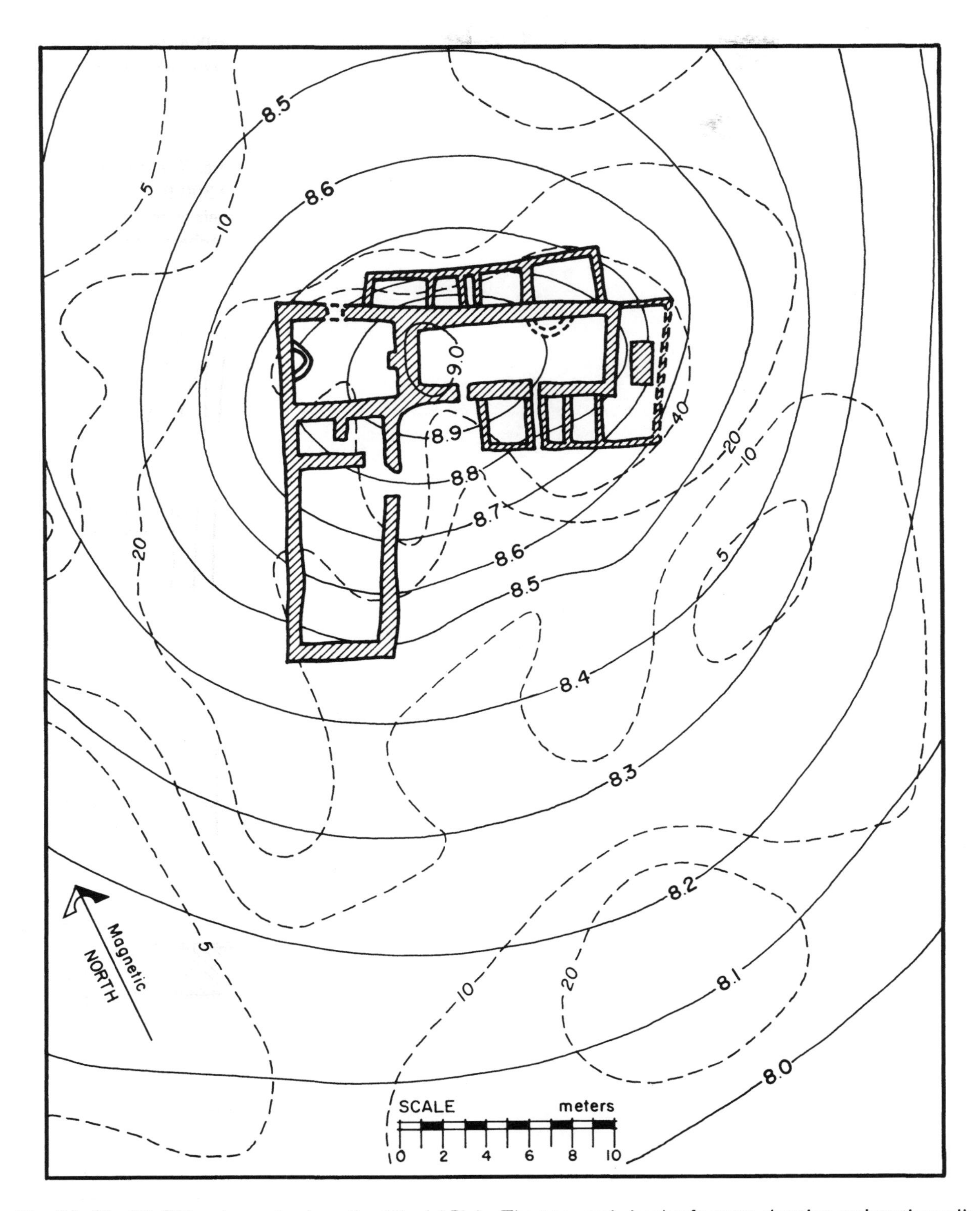


Fig. 3.4 Site DL-241, a homestead on the Alluvial Plain. The two semi-circular features abutting against the walls of two of the larger rooms are hearths. The rectangular feature in the easternmost room appears to be a ceramic box, perhaps a coffin. The solid lines represent contours in meters relative to an arbitrary datum. The dashed lines represent the relative densities of artifactual materials – the number of sherds and pieces of chipped stone per square meter.

that the Plain is undergoing a period of downcutting with little alluviation taking place (Kirkby and Kirkby 1969: 2-3), all were factors involved. From these types of data the following information has been derived in regard to trends in the site (community) and settlement patterns, as well as the water-control and irrigation systems utilized during the periods dealt with in this paper.

## ACHAEMENIAN, SELUCID AND PARTHIAN PERIODS (ca. 530 B.C. to A.D. 226)

The eleven sites illustrated in Figure 3.5 apparently all have an occupation spanning the Achaemenian through Parthian Periods. The site clusterings are consistent with the locations of major canal systems branching off the Mehmeh and Dawairij rivers at points in the approximate center of their courses through the plain. In these locations, the channels of the rivers range from about five to ten meters below plain level. However, since the channels are wide and have several terraces leading up from the water level to the plain, the excavation of a canal would not require the tremendous amounts of soil removal necessary further upstream. Upstream, the river channels are somewhat deeper and narrower, and are characterized by nearly vertical banks.

I was not able to recognize any unusual or distinctive features of the Achaemenian through Parthian canal systems, although a more intensive study with test excavations would undoubtedly provide a good deal of information. Systems of this time range are as yet poorly understood throughout Persia and Mesopotamia. The canals run from the midpoint of the river's course through the plain to take advantage of the terraces while providing the maximum amount of irrigated land. The length of these canals is difficult to determine: the Mehmeh systems

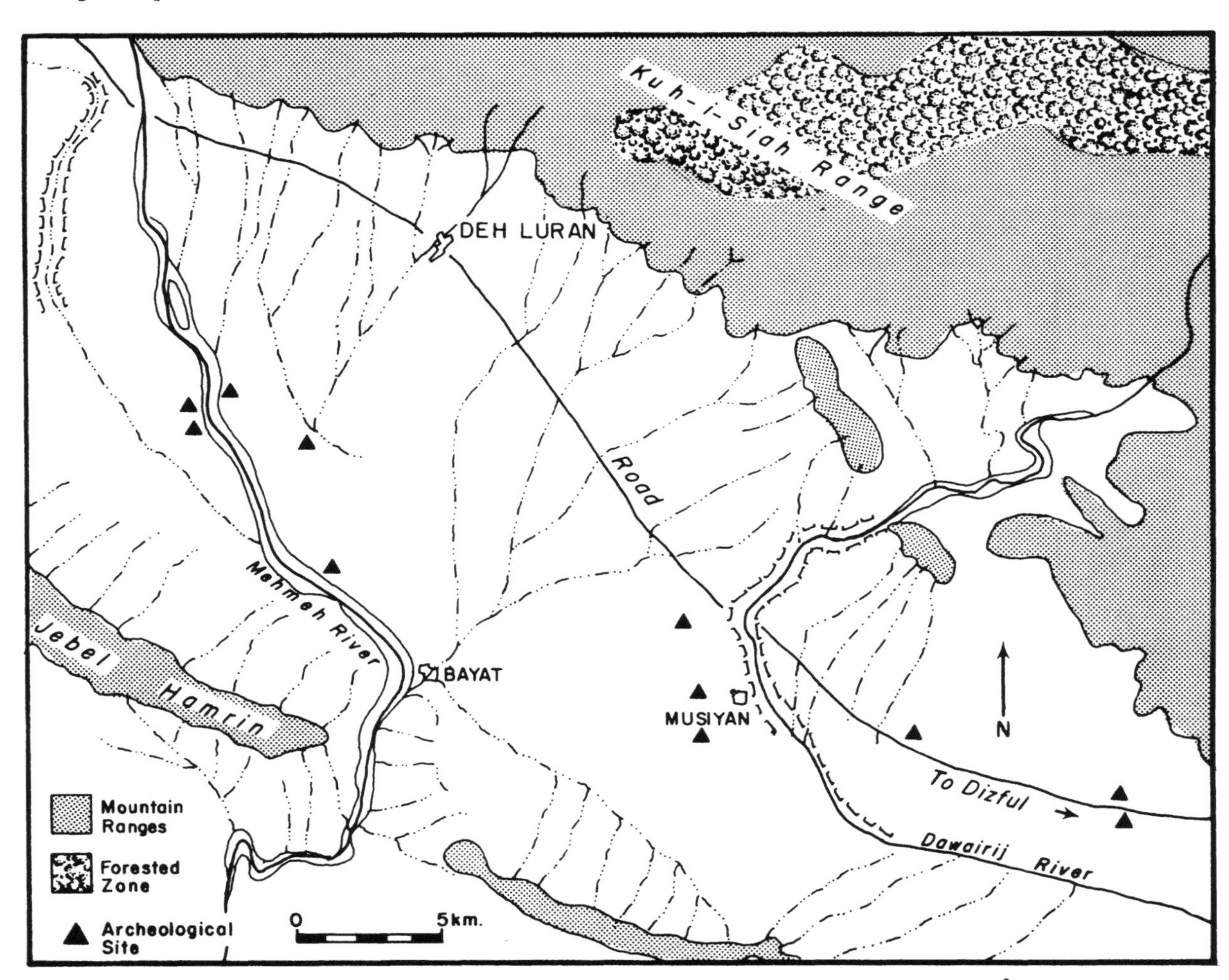


Fig. 3.5 The Deh Luran Plain, illustrating the site distribution (settlement pattern) for the Achaemenian through Parthian periods (ca. 530 B.C. to A.D. 226).

are about six kilometers and the Dawairij systems some fifteen kilometers. The canals are fairly large, ranging from approximately three meters to twenty meters in width, with an average width of about ten meters. There is a striking absence of sites in the Piedmont Zone (north of the improved road passing southeast-to-northwest through the Deh Luran Plain) and the central portions of the alluvial plain zone.

All of the eleven sites are quite large, between five and forty hectares, with a mean area of about sixteen hectares. They are all true tells (mounds), and at least three of the largest had walls surrounding them. Using Adams' (1965: 123) population estimate of about two hundred persons per hectare of built-up town or city, one arrives at a total population of about thirty-five thousand, or thirty-five persons per square kilometer for the area surveyed.

## SASSANIAN AND EARLY ISLAMIC PERIODS (ca. A.D. 226 to 800)

The Sassanian and Early Islamic settlement pattern (Fig. 3.6) is dramatically different from the preceding situation. The sites of these two periods have been presented together due to the difficulties in making a clear distinction between the Sassanian and Early Islamic ceramics in the Deh Luran Plain, and because the vast majority of the 249 sites apparently were occupied continuously through both periods.

The clusterings of sites were consistent with the locations of newly introduced water-control and irrigation techniques in the upper reaches of the Mehmeh and Dawairij rivers, and the Piedmont Zone. The canal systems supplying water to the sites in the lower Mehmeh and Dawairij area, where we see quite

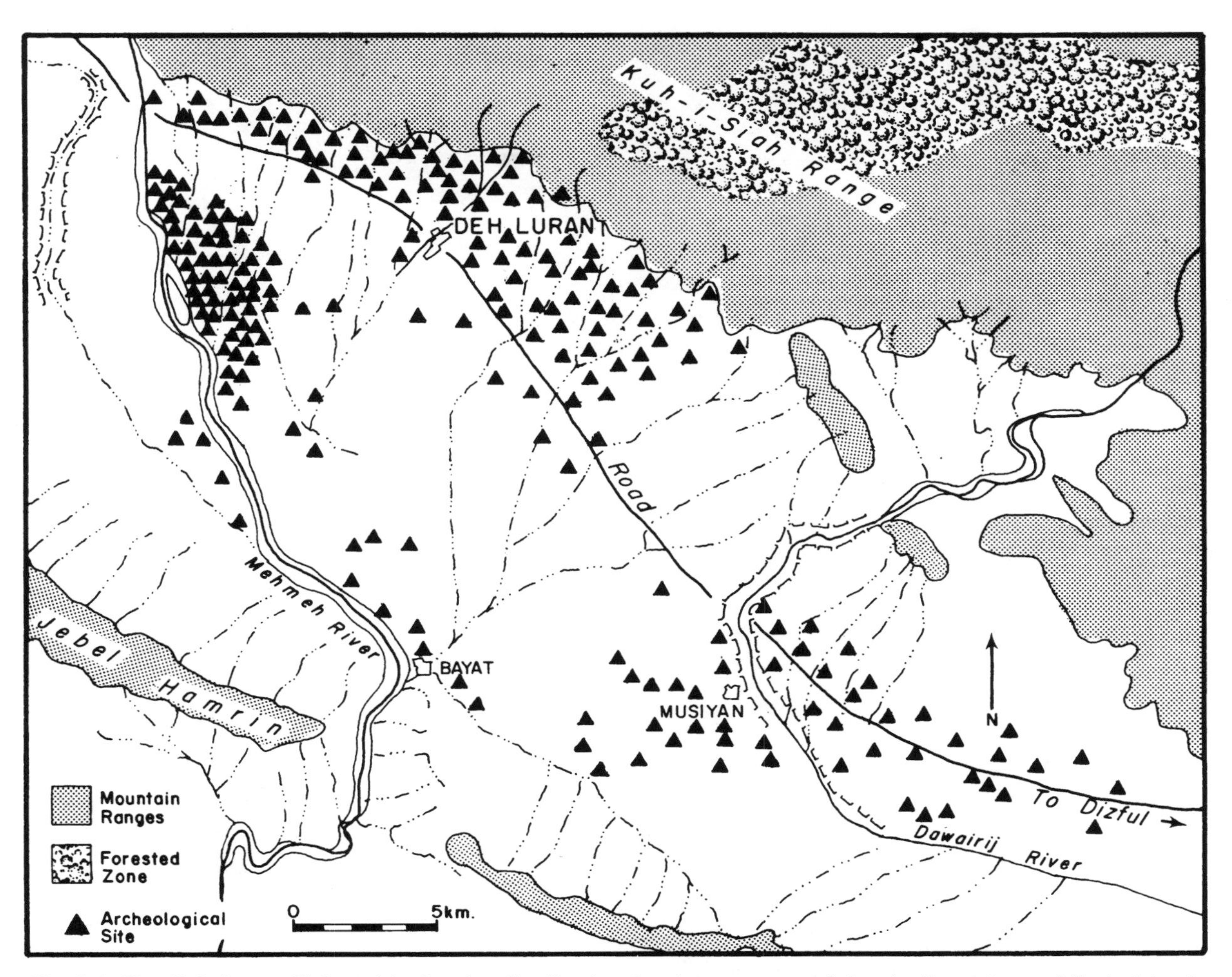


Fig. 3.6 The Deh Luran Plain, with the site distribution (settlement pattern) for the Sassanian and Early Islamic periods (ca. A.D. 226 to 800). Because the scale is too small, this figure does not show all 249 sites attributed to these periods. However, the locations and boundaries of the site clusters are correct.

a large increase in the number of sites, were not entirely new and, as far as I can tell, only in one instance was a technological innovation (a qanat system on the lower Dawairij) utilized. However, additional canals were excavated to service new sites and further increase the water supplies for irrigation in the lower area.

### Upper Reaches of the Mehmeh and Dawairij Rivers

The major irrigation method newly introduced in the northern half of the Deh Luran Plain was, in reality, an ingenious modification of the qanat technique. Qanats, apparently first utilized elsewhere in Pre-Achaemenian times, are slightly inclined tunnels excavated from a water source or aquifer toward an area to be serviced. These tunnels are connected with the surface of the earth by a series of regularly spaced vertical shafts, which serve in the removal of debris during excavation and cleaning. The gradient of the qanats, gentler than the surface slope, is designed in such a way as to intersect the surface gradient as near as possible to the cultivation. From the point of intersection, surface channels take water to the fields. The reader is directed to articles by Cressey (1958), English (1968), and Wulff (1968) for further details on qanats throughout the Middle East.

The Deh Luran systems, built in the Sassanian and Early Islamic Periods, entailed the modification of the qanat principle to tap seepage of the Mehmeh and Dawairij where their channels were deepest and their banks most vertical. These qanats did not take water directly from the rivers, but ran parallel to them at a distance of from ten to fifty meters for stretches of from two hundred meters to nearly two kilometers. The qanats then usually turned in toward the center of the alluvial plain. Apparently water was obtained from the rivers as it percolated through the soil and rock of the banks into the qanat systems. This modification of an already existing technique evidently was contrived to serve two purposes. First, it greatly reduced the amount of silt carried into the systems – a problem that must have been enormous in the maintenance of the canal systems that took water directly from the Mehmeh and Dawairij rivers by means of diversion dams or weirs. Secondly, the percolation of the water through the river banks filtered out vegetal matter and very probably minerals in suspension, thereby reducing the need to clean the systems and slowing down the salinization process that was gradually affecting crop production.

In this same general area, but extending toward the north-central portion of the plain, a second, less impressive technique was also evidently introduced at this time. This technique consisted of numerous low terraces and check dams constructed in association with many of the intermittent drainages dissecting the plain. At first glance these terraces and check dams appear to have functioned in a manner similar to the larger scale dry-farming features to be described below for the Piedmont Zone. However, it seems more likely that they functioned in combination with more usual irrigation techniques to make available land otherwise lost to cultivation.

## The Piedmont Zone

With the Sassanian introduction of new techniques, the piedmont microenvironment was converted from a largely uninhabited zone to one of rather dense occupation. As illustrated in Figure 3.6, this occupation was focused in the western half of the piedmont and consisted of eighty-nine sites. This type of occupation is in sharp contrast to the two relatively small and short-lived sites that represented the seven thousand years of Pre-Sassanian occupation of this zone.

*Dry-Farming Water-Control Systems.* In the piedmont the Sassanians utilized a technique to conserve and better allocate rainwater rather than bring water to the area from a source such as a river or spring by means of an irrigation system of canals and qanats. This control was accomplished through the construction of low terrace walls of unmodified boulders and cobbles. The terrace walls were built at intervals varying with the slope of the terrain, wider apart in areas of less slope and closer together as the slope increased, generally following the contours of the land. These walls functioned to retain soils and runoff waters from higher terrain, thereby permitting a more efficient use of the rocky slopes for seasonal dry-farming cultivation.

An integral part of this system was a series of small check dams of boulders, constructed at right angles to the flow across the intermittent washes or drainages dissecting the piedmont. Water and soils seeping through the terrace walls, as well as coming from unterraced areas, would find their way to the natural drainages. The check dams functioned to retain soils and retard runoff so that the water would thoroughly soak these soils. The small plots behind the check dams tended to have relatively deep soil deposits. Because of these deposits, and the fact that the plots received greater amounts of rejuvenating soils and water than the slope terraces, the plots probably produced greater quantities of cultigens with a higher degree of reliability. Because of their strategic locations, the plots and the dams were more likely to be destroyed during violent thundershowers, and therefore required more careful maintenance.

The piedmont agricultural terraces and check dams tend to cluster in distinct units in association with one or more habitation structures (usually having a walled courtyard), one or more small "storage" structures, and frequently one or more square or rectangular structures reminiscent of cattle pens or "corrals." Figure 3.7 is a plane-table map of site DL-194, and illustrates a classic example of one of these units. I propose that these are distinct socioeconomic units – dry-farming homesteads. My rationale for this proposal is the fact that each such unit of houses, corrals, farming terraces, and check dams is clearly separated by some twenty-five to one hundred meters from the next. In many cases this separation is further delineated by an intermittent drainage, containing no check dams, or by a low wall of unmodified boulders. These walls, such as that illustrated in the southeast quadrant (Fig. 3.7), may well have

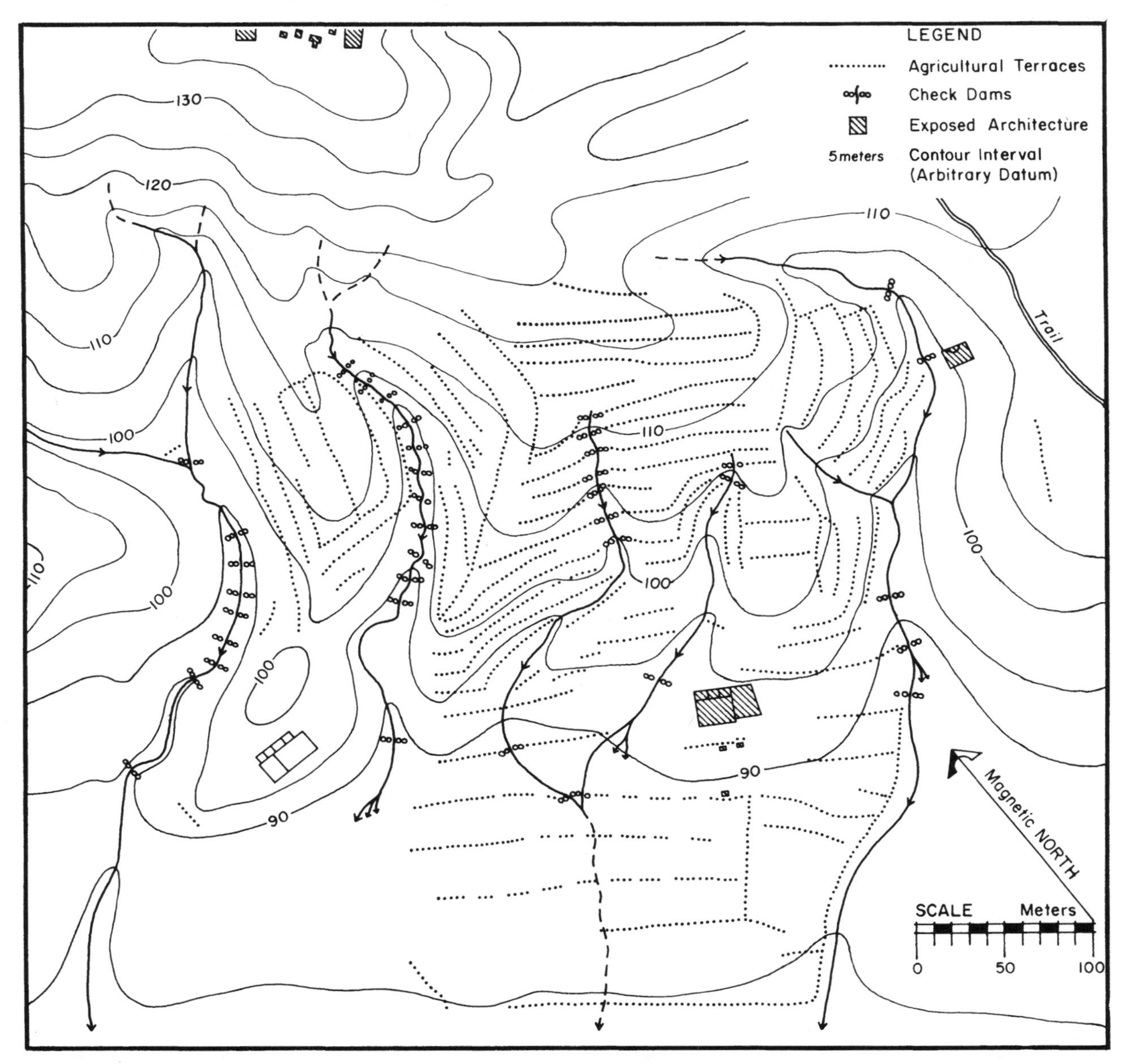


Fig. 3.7 Site DL-194, a dry-farming homestead in the Piedmont Zone. The structures illustrated in the northern extreme of this map form part of a hilltop site. Hilltop sites, distributed one for every four to six homesteads, may have served as defensible refuges for the agriculturalists.

served the dual function of delineating the boundary and serving as an agricultural terrace wall.

Through these data we may derive information beyond the immediately apparent material culture. For instance, we may determine the ratio of habitation area and population to the area of cultivation, and thereby obtain some idea of the area needed to sustain a specific size of population in the piedmont zone through the use of dry-farming techniques. It is, of course, recognized that such calculations are very tentative due to the problems in estimating population and our poor knowledge of Sassanian-Early Islamic farming techniques and land tenure. Another factor difficult to assess is the scale of dependence on domesticated cattle. Were the piedmont homesteads, utilizing seasonal dry-farming techniques, able to derive a surplus for storage or trade to see them through the summer months? Did they base their economy partially on cattle, as may be indicated by the presence of corral-like structures? Was such supplementary cattle implied by the existence of low walls with no apparent function other than to divide sections of land within a large, apparently uninhabited area of perennial grasses in the eastern piedmont? Would the environment permit such a sedentary livelihood? An alternative explanation would be that the piedmont homesteads were occupied by a transhumant pastoral group or groups, living on the Deh Luran Plain only during the winter. As noted previously (Hole, Flannery, and Neely 1969: 349-50), we feel there is a distinct possibility that the transhumant round practiced today by the Lurish nomads between the Khorrambad Valley and Deh Luran Plain extends back in time perhaps as far as 6700 B.C.

In any case, using site DL-194 as an example of a piedmont homestead, we find the total living area (including the walled courtyards) of the two structures to be 1,088 square meters, or 289 square meters if we consider only the area of the fifteen to seventeen definable rooms. The ratio of total living area (including courtyards) to the area under dry-farming cultivation at DL-194 (totaling 16.88 hectares) is 1:153 square meters.

Evidently structures of this type and size are not usual in the comparative ethnographic data provided by Adams (1965: 23-25). However, since those are the best data presently available, we might estimate a homestead population of between ten and twenty-nine individuals. These two figures represent the extremes, and were calculated assuming that both structures were occupied contemporaneously and that half of the fifteen to seventeen rooms were used as living quarters. The relatively large population size indicated for this homestead would perhaps suggest a social unit definable as an extended or joint family. These data beg for further investigation. The final analysis of all the piedmont sites in Deh Luran may show that the habitation structures at DL-194 were significantly larger than the mean, or that both structures were not inhabited contemporaneously. Special efforts should be made to search Sassanian and Islamic documents and inscriptions for information bearing on these problems.

An interesting comparison can be made between the preceding settlement type and contemporary homesteads or hamlet units based on irrigation agriculture. These sites were found on the alluvium near the Mehmeh River in the southwestern quadrant of the Deh Luran Plain. Figure 3.8 illustrates two of these units, DL-274 and DL-275, composed of small, low house mounds and associated fields irrigated by canals. The units were divided from one another by large secondary canals which derived

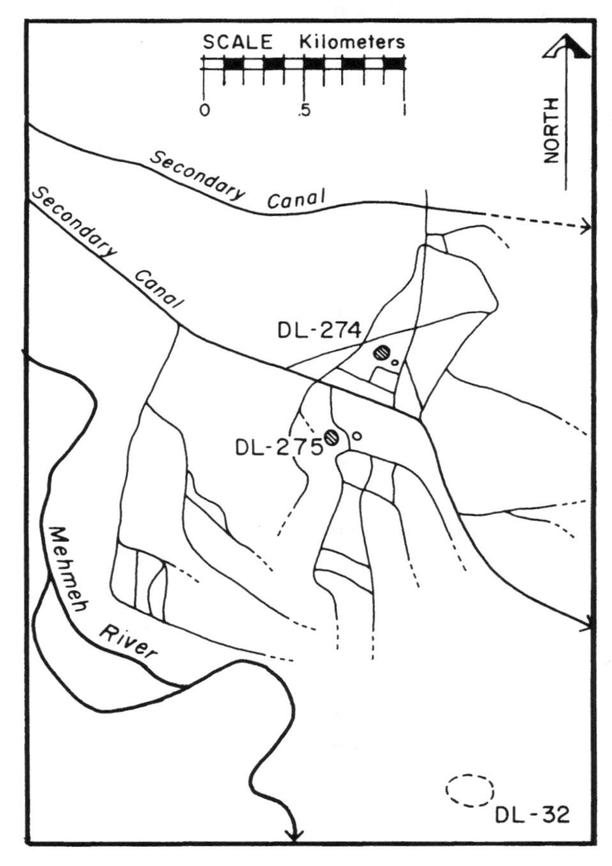


Fig. 3.8 Sites DL-274 and DL-275, homesteads on the Alluvial Plain.

water from main canal and qanat systems. Smaller feeder canals branched off the secondary canals to form a dendritic pattern in the cultivated fields of each homestead or hamlet unit. The secondary canals functioned as parts of a larger system that apparently supplied water to numerous such homestead or hamlet units along their courses, and probably also drained excess water from the fields to retard the buildup of salts. The water passing down the northernmost secondary canal (Fig. 3.8) supplied site DL-274. Excess water drained into the other major canal to be utilized by site DL-275 or passed to the next homestead or hamlet unit southeast. This system seems to have ultimately drained into the salty marshes lying to the southeast. However, the presence of poorly defined segments of major canals in the southwestern extreme of the survey area indicates that these waters may have drained into a large canal system that ran southeast from the Mehmeh River, to supply water to the southern segment of the plain. A very important function of this huge southern canal system may have been to drain the salty marshes in order to retard salt buildup and possibly to reclaim some of the marsh for agricultural use.

In any case, to return to the comparison with piedmont sites, using DL-274 and DL-275 as examples of alluvium homesteads or hamlets, we find the total living area of the two sites (the house mounds and surrounding area having building debris and a relatively dense sherd scatter) to be 7,174.3 square meters and 9,190.2 square meters respectively. Population estimates are even more specious in these examples since the exact number of structures, rooms, and their sizes are not known. However, given the area involved, I would hazard an estimate of 60 to 180 persons for DL-274 and 80 to 240 persons for DL-275. The ratio of total living area to the area under irrigated cultivation is 1:173 square meters for DL-274 and 1:220 square meters for DL-275.

*Spring-Fed Canals with Drop Tower-Mills (Arubah Penstocks).* Another Sassanian technical innovation in the piedmont was a system of canals (DL-5) fed by the Ab-i-Garm springs. These small, closely associated springs are located high in the piedmont near its juncture with the foothills of the Kuh-i-Siah. While there are a number of springs similarly located along this break in the topography, evidently only the Ab-i-Garm sources, north-northeast of the present town of Deh Luran, were canalized in the manner to be described.

The spring waters were diverted into the canal system either at the spring heads or just downstream (Fig. 3.9). There was no evidence as to exactly how this diversion was accomplished, but a type of diversion dam seems most likely. The relatively small, broad, U-shaped canals, averaging 1.5 meters wide by 0.6 meters deep, were constructed into both faces of the banks (Fig. 3.10) of the natural drainage that carried Ab-i-Garm waters down the slope to the alluvial plain. The canals carefully followed the contours of the banks to retain a very gentle, but efficient grade. Where subsidiary or secondary natural drainage features cutting the Ab-i-Garm bed were particularly broad or deep, and would thereby require the excavation and maintenance of many meters of additional canal to follow the contours properly, an aqueduct was constructed to span the secondary drainage. One of the better preserved aqueducts was excavated in the northern portion of the system (Fig. 3.9).

In the upper (northern) 2.25 kilometers of the system, well-made, tower-like structures of mortared masonry were built against and partially into the steep face of the drainage banks (Fig. 3.10). These tower-like structures were constructed at irregular intervals ranging from approximately 50 to 600 meters apart, with an average distance of about 275 meters separating these features. The towers received the canalized water and dropped it some 6.5 meters into a continuation of the canal system.

The remaining 4.25 kilometers of the system, in which the canal was carried southward while the natural Ab-i-Garm drainage changed course toward the southwest (Fig. 3.9), involved a more laborious method of construction. Here, trenches were excavated into the piedmont slope so as to duplicate the gently graded canals present in the upper segment of the system. As this canal grade was more gentle than that of the piedmont, the trenches excavated were necessarily deeper at their upper (northern) end and became proportionately more shallow as they coursed southward. The towers were set into the trenches at their deep, northern extremes. This portion of the canal system probably was built in response to the domestic water needs of the communities (DL-3, DL-4, and DL-312) through which it passed as well as the irrigation needs of fields located on the alluvium immediately south of the piedmont zone. Prior to the completion of the system, the major part of the alluvial plain which it watered could have been useful only for seasonal dry farming and grazing.

Large sections of the system were in poor condition due to natural erosion, localized alluviation, stone robbery for reuse in modern structures, and the

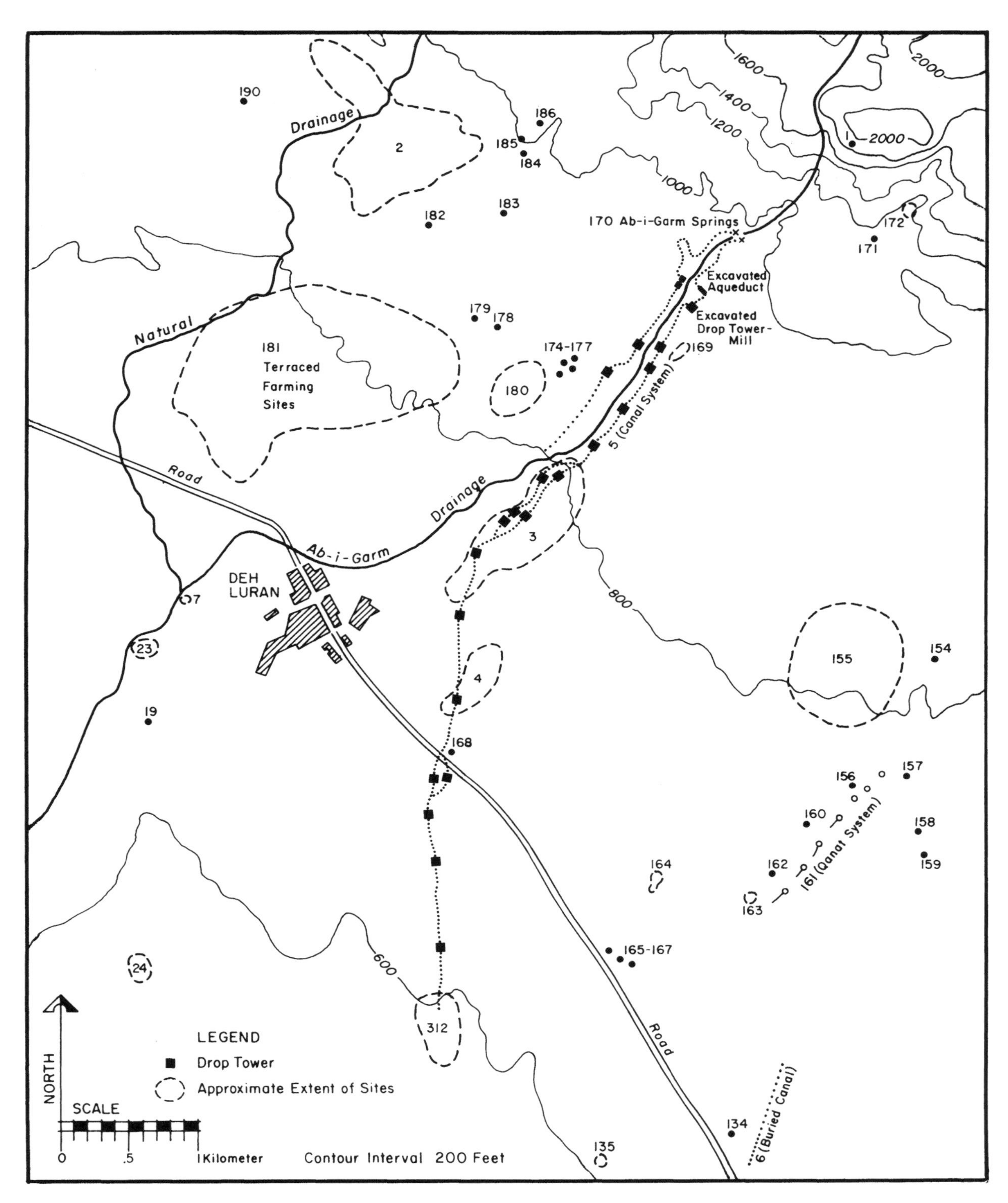


Fig. 3.9 The north-central portion of the Deh Luran Plain. Note the details of the canal system (DL-5) with drop tower-mills.

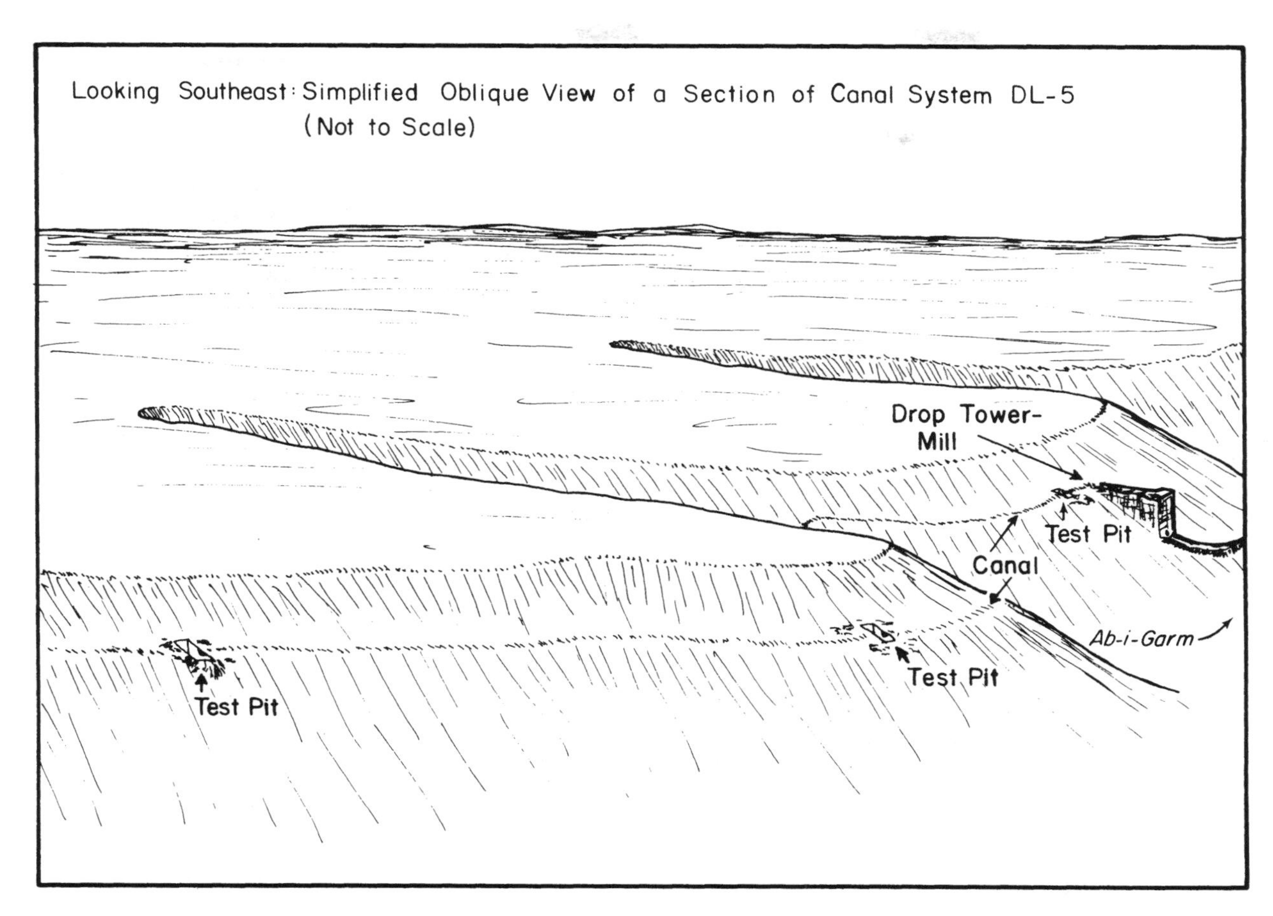


Fig. 3.10 A segment of canal system DL-5 nears its point of origin. A line of lush grasses within a very shallow depression indicates the presence of the canal. The bed of the Ab-i-Garm drainage lies in front and to the right of the drop tower-mill.

filling-in or covering-over of the canals and towers-in-trenches by earth and debris resulting form modern agriculture. These factors made it impossible to determine the original length of the system and the total number of drop-towers. The system was traceable for a straight-line distance of about 6.5 kilometers, within which the remnants of 22 drop-towers were found.

Upon first seeing these drop-towers, I interpreted their function as correlated with the efficiency of the canals in relation to the slope of the piedmont. That is, as the gradient of the piedmont was too great to allow the construction of the more gently sloping canal, the towers served to drop the water an appreciable distance into a continuation of the system. In this manner, so I thought, an efficient water flow was attained from the spring heads to the alluvial plain through the series of step-like segments of the system.

However, this first impression was soon modified due to the excavation of one of the towers and several small test trenches which cut through the canals at right-angles to the flow (Fig. 3.10), as well as a reconsideration of the gradients involved. The test trench excavations permitted an accurate measurement of the gradient for the segment of the canal located between the excavated aqueduct and drop-tower (Fig. 3.9). In this part of the system the gradient of the canal is very gentle – ranging from approximately 0°20′ to 0°35′ (or about 0.6 to 1.02 percent), with a fall of some 3 centimeters every 30 meters (0°30′ or 0.87 percent) about average. In measuring the gradient of the piedmont, I found its slope more gentle than I had thought. Topographic maps (Fig. 3.9) indicate a drop of approximately 400 meters in a distance of 6.5 kilometers, which is a grade of 1:53, about 1 degree or 1.8 percent. Later research revealed that an ordinary canal excavated into the piedmont, with some planning to avoid major topographic irregularities, would have functioned nearly as efficiently as the canal and drop-tower system. For example, a general discussion of irrigation in arid lands states that "irrigation by gravity-flow through water channels and over the ground surface is difficult to maintain on slope exceeding 2° (3.5%), and very difficult on slopes exceeding 4° (7.0%)" (Greene 1966:258). Current advice provided by the U.S. Soil Conservation Service (Department of Agriculture 1959:20) to farmers in the western United States, however, suggests that flatter grades are more efficient:

> You should lay out your ditches on relatively flat grades, having a fall of not more than one or two inches per hundred feet (0.8% pr 1.7%). Ditches with a fall of more than three inches per hundred feet (2.5%) are apt to erode because of the higher flow velocities.

If, then, the natural gradient of the piedmont was not the reason behind the construction of the towers, what was? The excavation of one of the towers (Fig. 3.11) in the northern portion of the system (Fig. 3.9) revealed architectural and artifactual materials to indicate that these towers provided water power to drive mill wheels for grinding grains. This drop tower-mill construction is technically termed the "*Arubah* penstock." Avitsur (1960) illustrates a similar tower-mill and discusses its functional qualities in some detail.

## Settlement and Site Patterns

In the settlement pattern for the Sassanian and Early Islamic periods on the Deh Luran Plain (Fig. 3.6), the sites tend to be smaller, more numerous, and much more dispersed than in the previous periods. The sites vary greatly in size from one family farm houses (e.g., Fig. 3.4), some of which were less than 100 square meters in floor area, to urban centers of which at least one is nearly 100 hectares.

The mean total area of the sites dating to these periods is about 1.5 hectares. While this size is a

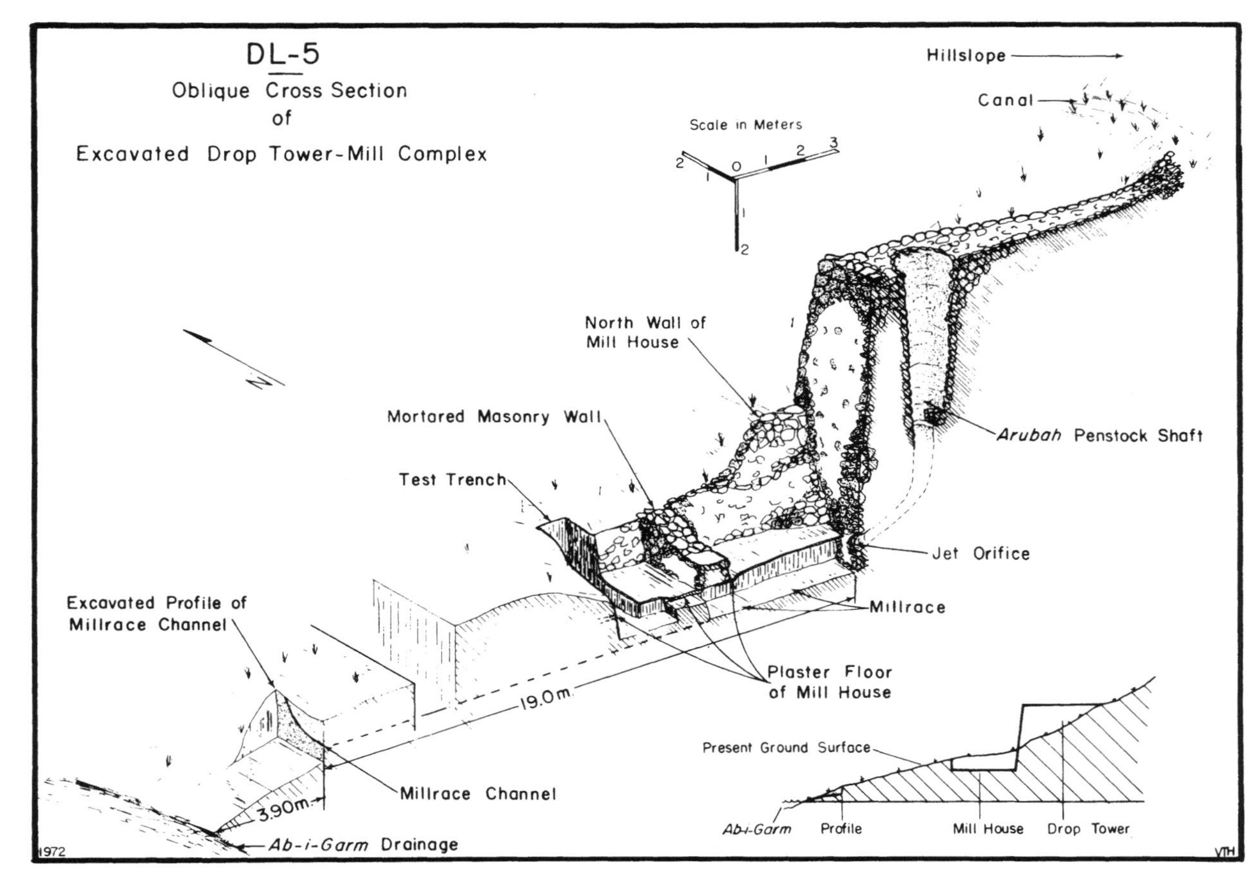


Fig. 3.11 Cross-section of the excavated drop tower-mill complex, which forms an integral part of the canal system DL-5 near its point of origin at the Ab-i-Garm springs.

dramatic reduction from the mean total area of some sixteen hectares for sites of the Parthian period, the increase in the total number of apparently contemporaneously occupied sites (from 11 to 249) indicates an increase in population. I would estimate the population of the Deh Luran Plain for the Sassanian-Early Islamic periods as nearly 75,000 individuals, or some 75 persons per square kilometer for the area surveyed. While large, this figure is apparently consistent with the carrying capacity of the land – but probably just barely so.

The areal settlement pattern displays a lineality, with strings of sites located alongside canals. The pattern of individual sites is likewise linear. Sites are long and narrow, with units (households) dotted in a band for as many as hundreds of meters along both sides of a canal. Construction features within the larger sites are much more dispersed than in previous periods. Structures consist of low mounds and platforms of earth and brick, as well as foundations of stone slabs and boulders that today lie half buried in the alluvium and clay eroded from upper walls.

In summary, the addition of new water-control and irrigation systems, and the enlargement of the older, pre-existing systems permitted the inhabitants of the Deh Luran Plain to occupy and utilize an additional surface area of some four thousand square kilometers, or approximately 40 percent of the total area of the Plain. For the ten thousand years of habitation in the Plain, the Sassanian and Early Islamic periods are the peak in terms of total amount and intensity of land use as well as maximum population density.

## LATER ISLAMIC PERIODS (A.D. 800 to 1250)

It is beyond the purpose and scope of this paper to delve into later periods of occupation on the Deh Luran Plain. However, as with the Achaemenian through Parthian periods, I would like to present for comparison the trends in the utilization of irrigation resources and settlement and site patterns for the two subsequent Islamic periods of A.D. 800 to 950 and A.D. 950 to 1250.

### A.D. 800 to 950

Apparently as the Sassanian and Early Islamic periods witnessed dramatic modifications in agricultural technology and settlement, the subsequent Islamic Period of A.D. 800 to 950 saw yet another set of changes. While the individual site pattern remained essentially unchanged from the Sassanian-Early Islamic periods, distinct changes in the settlement pattern did occur (Fig. 3.12). The modifications were once again consistent with the locations of major water-control and irrigation systems functioning at the time.

The types of water-control and irrigation systems were essentially the same as those of the Sassanian-Early Islamic periods. However, the number and extent of these systems were severely reduced, and apparently the construction of new systems or additions to the existing systems was completely curtailed. This reduction was especially evident throughout the Piedmont Zone, along the northern edge of the Alluvial Plain Zone, and along the northern reaches of the Mehmeh River.

In the Piedmont Zone only a few of the dry-farming homesteads continued to be used, and it seems likely that the canal with drop tower-mill system was only operating on a limited basis, if at all. There is some evidence that the present-day portion of the Ab-i-Garm drainage by means of canals and a form of floodwater irrigation may have originated in this period. Cultivation of this type could easily have been practiced by the inhabitants of the three sites found very near the modern town of Deh Luran (Fig. 3.12). The survey did not find a sufficient amount of associated cultural materials to date with any accuracy the periods of use of these terraces. In fact, the utilization of these terraces may well date from the early portion of the Uruk Period (ca. 3,500 B.C.), and thus explain the presence of site DL-169, one of the two sites located on the piedmont prior to the Sassanian-Early Islamic periods. Irrigation agriculture along the northern edge of the Alluvial Plain and the northern reaches of the Mehmeh River apparently ceased to exist with the abandonment of the qanat systems and the canal with drop tower-mill system that supplied the necessary water.

The reduction in the total number and scale of water-control and irrigation systems was accompanied by a reduction in the number of sites (from 249 to 20) occupied during this period. These sites varied greatly in size, but the mean area tended to be larger than in the Sassanian and Early Islamic periods. The extremes in size range from approximately 1,000 square meters (0.1 hectare) to about 90 hectares, with a mean area of about 12 hectares for the 20 sites recorded. These data suggest a population of about 48,000 persons, or some 48 individuals per square kilometer for the area surveyed.

## A.D. 950 to 1250

The trends toward a less intensive land use and lighter population density indicated in the previous period were maintained until the apparent abandonment of the Deh Luran Plain at about A.D. 1250. The dispersed pattern within individual sites, present since Sassanian times, remained essentially unchanged. However, the overall settlement pattern underwent modification (Fig. 3.13). The changes are positively correlated with the water-control and irrigation systems functioning at the time.

Techniques of water control for dry farming probably were used only around the relatively large site of DL-2, located on the piedmont to the northwest of the modern town of Deh Luran. Water for domestic use, and perhaps some very limited irrigation by the inhabitants of DL-2 was probably obtained from the small springs located in the topographic-geological break between the piedmont and the foothills of the Kuh-i-Siah at the northern extreme of the site.

The two sites southwest of the modern town of Deh Luran very probably diverted waters from the Ab-i-Garm drainage to irrigate the alluvial terraces and parts of the alluvial plain adjacent to the watercourse. Ceramics of this period were collected from the fields surrounding these sites, but the actual boundaries and characteristics of the fields, as well as the location of the associated irrigation canals, were not identifiable.

It seems likely that the inhabitants of the plain ceased to use the Mehmeh River as a source for irrigation at the end of the previous period. This cessation is indicated by the lack of evidence for occupation in the western portion of the plain, and the fact that no ceramic materials dating to this period were found in association with irrigation features originating from the Mehmeh. This abandonment may well have been at least partially caused by the more salty or brackish nature of the Mehmeh River. It is probable that crop production would have been somewhat greater through the use of waters

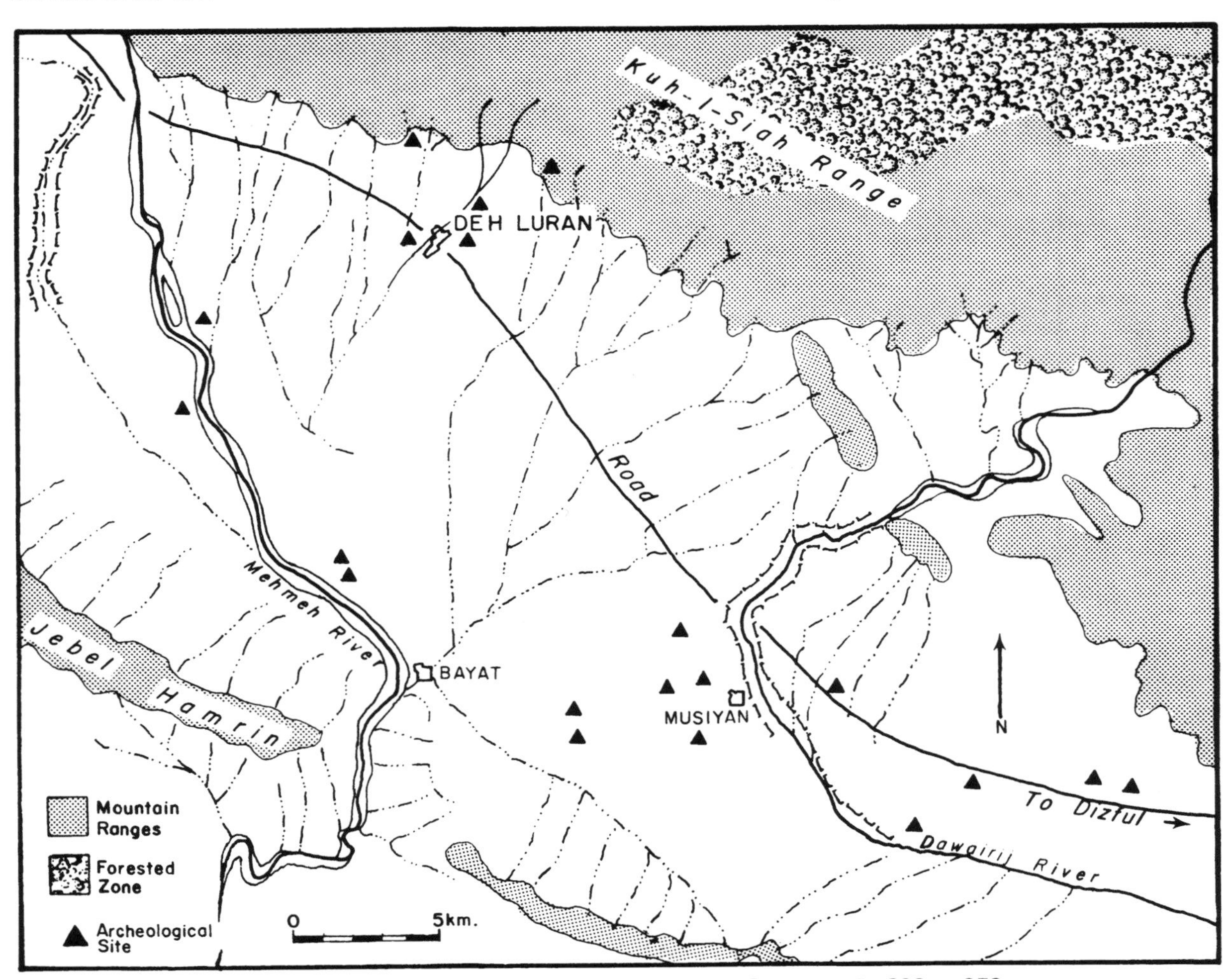


Fig. 3.12 The Deh Luran Plain, with site distribution for the Islamic Period, A.D. 800 to 950.

from the Dawairij River. Unfortunately, at this point in the study of the occupation of the Deh Luran Plain, it is not clear if the dwindling population resulted in the reduction or curtailment of irrigation systems or vice versa. I suspect both. That is, a reinforcing reciprocal process was actually involved.

The sites located in the east-central portion of the plain were serviced with domestic and irrigation waters from the Dawairij River by means of two fairly large qanat-canal systems. One of these systems carried water south and west of its take-off point from the river, while the second coursed toward the southeast.

A reduction in the number of sites (from twenty to eight) is apparently correlated with the reduction in size and the use of fewer water-control and irrigation systems during this period. The mean total area represented by these eight sites amounts to about fifteen hectares, with the extremes of size ranging from approximately five thousand square meters (0.5 hectare) to about sixty hectares. A population of some twenty-four thousand individuals, or about twenty-four persons per square kilometer, may be estimated for the area surveyed.

## CONCLUSIONS

The pattern of population increase and a change in the settlement pattern accompanied by concomitant innovations and expansions of the water-control and irrigation systems during the Sassanian Period is not unique in the Deh Luran Plain. Adams (1965:69-83, 116-17) has found much the same pattern existing in the Diyala region of Mesopotamia to the west, and on the Upper Khuzistan Plain to the east. There are, however, certain differences in the Deh Luran pattern that suggest new information to augment our very sketchy knowledge of the Sassanian sociopolitical organization and the processes of related developments and changes that took place at the beginning and termination of the Sassanian Period.

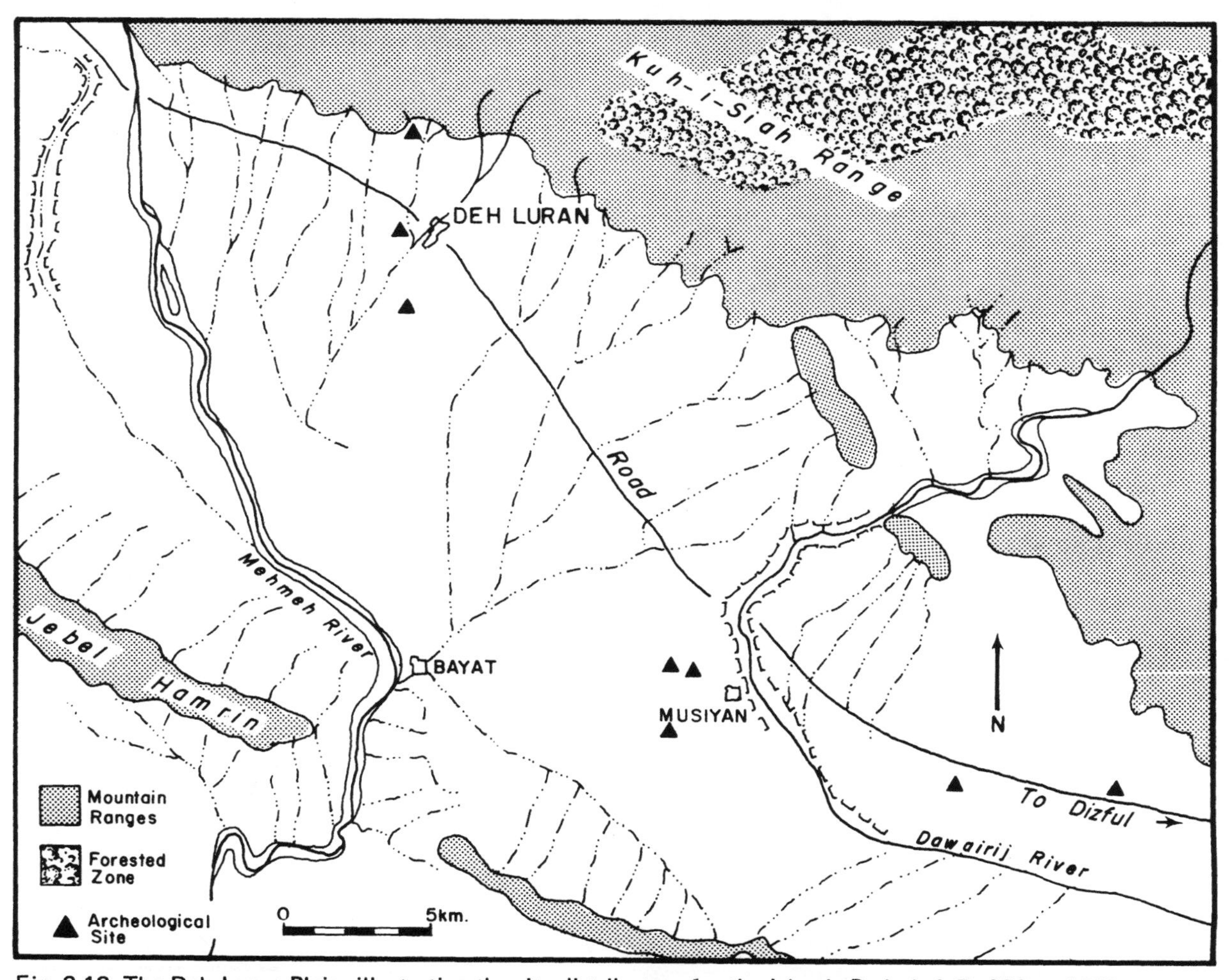


Fig. 3.13 The Deh Luran Plain, illustrating the site distribution for the Islamic Period, A.D. 950 to 1250.

## Water-Control and Irrigation Systems

The most basic of these differences are the technological introductions in the form of water-control and irrigation techniques and features. Each of the techniques, apparently first introduced to the Deh Luran Plain during the Sassanian Period, had precursors or contemporary counterparts in other parts of the Middle East. The use of qanat systems has been indicated by early documents to have existed probably as early as the seventh century B.C. when the Assyrian king Sargon II (722 to 705 B.C.) was recorded as having destroyed such a system around the town of Ulhu near Lake Urmia, presently in the province of Azerbaijan in northwestern Iran (Forbes 1955:152-53). Qanat systems were almost certainly present in both Mesopotamia (Iraq) and Persia (Iran) by the sixth and fifth centuries B.C. (Forbes 1955:154-59). The construction of dry-farming terraces and check dams can be traced back probably to as early as the tenth century B.C. (Evenari et al. 1961:981-82), and surely to the second century B.C. (Kedar 1957a, 1957b; Colt 1962:13-17; and Mayerson, 1962) in the Negev region of Palestine. Personal observations and conversations with colleagues have indicated the presence of drop tower-mills (*Arubah* penstocks) throughout much of Iran.* Drop tower-mills, very similar to those found in Deh Luran, have been documented for perhaps as early as the second century B.C. in Phoenicia and almost certainly for the fourth or fifth century A.D. in Palestine, Lebanon, and southern Syria (Avitsur 1960:39-41).

To the best of my knowledge, the exact forms or modifications of the techniques and features that characterized the Deh Luran Plain during the Sassanian-Early Islamic periods have not as yet been reported from other locales in the Middle East. There is, of course, no reason to believe Deh Luran unique in this respect. With our interest now shifting to the discovery and definition of such techniques and features, our fund of knowledge increasing as to the forms of the resultant systems as well as their interrelationships with certain topographic and environmental conditions, and the increase in the number of specific problem-oriented, multidisciplinary studies being initiated toward the investigation of such systems, it seems likely that similarities will be found elsewhere.

The important factor I wish to stress here, however, is not the uniqueness of the systems, but the relation of each system to the microenvironmental situation in which it was utilized, and the impact of their introduction on the society and culture of the area. Because of the earlier history of use in other parts of the Middle East, I prefer to look on the ideas underlying the techniques, features, and resulting systems as having been brought into the Deh Luran region and individually modified to the particular topographic and environmental situations at hand. I view these modifications as the work of experienced engineers adapting known, tried-and-true techniques to the Deh Luran Plain in order to permit an increase in crop production through the utilization of previously little-occupied and cultivated portions of the plain. The innovations also would have allowed population increase. I will develop the bases of my contention as I discuss the major points in the following sections.

*Lynn Barry Fredlund 1969; and Frank Hole and Brian Spooner 1972: personal communications.

## Settlement and Site Patterns

There are four major points or observations that should be emphasized in relation to the settlement and site patterns. I will draw heavily on Adams' work in Khuzistan and Diyala for comparative material.

1) First, unlike the Diyala and Upper Khuzistan regions, (Adams 1965:63, 1962:115-16), there was apparently a general continuity of settlement and site patterns as well as population density during the Achaemenian, Selucid, and Parthian periods on the Deh Luran Plain. Why did the Deh Luran Plain remain more or less stable while the Diyala and Upper Khuzistan regions experienced an ongoing, and even somewhat dramatic, increase in the number and size of sites, and a concomitant growth in population density from the Achaemenian into the Parthian Period? I would propose that the continuity in Deh Luran involved two factors. (a) In contrast to its early history (Hole, Flannery, and Neely 1969), the Plain had, by this time, become a relatively marginal area, especially as far as urban development was concerned. Quantitative comparisons of site size, the number of sites in each size category, and settlement and population densities between Deh Luran and the Diyala Region best express this marginality. Adams' (1965:59-68) data from the Diyala indicate not only greater numbers proportional to the area surveyed, but the presence of categories of settlement (e.g., "small urban centers" and "cities") that apparently did not exist on the Deh Luran Plain prior to the Sassanian Period.[2] (b) Relative to its marginal status and technological developments (water control and irrigation), the Deh Luran Plain had probably reached a peak or equilibrium of population to available land and resources.

2) Second, the changes in settlement and site patterns at the onset of the Sassanian Period are apparently similar for the Diyala, Upper Khuzistan, and Deh Luran regions. Although they are rather difficult to determine, due to the different nature of the survey data available from these three regions, the patterns of land use and general population densities also appear to be similar. The growth in population and use of the available land and other resources appear to have reached an all-time maximum for the three regions during Sassanian times.

3) Third, while there evidently was a relatively rapid decrease in the number of settlements and the population density at the end of the Sassanian Period (ca. A.D. 637) in Diyala and Khuzistan (Adams 1965:81-111, 1962:119-20), Deh Luran appears to have been an exception. A decline does not appear to have taken place on the Deh Luran Plain until some time within the Islamic Period (ca. A.D. 800). The continuity is emphasized by the uninterrupted occupation of most of the Sassanian sites into Early Islamic times. Adams' survey data (1965:81-82, 1962:120) indicate that in the Diyala and Upper Khuzistan regions, up to half of the total settled area was abandoned at the end of the Sassanian Period and that new sites were founded by the early Islamic inhabitants. Quite often, the Early Islamic sites were located adjacent to, or very near, abandoned Sassanian settlements.

4) Fourth, the relatively rapid reduction in the number of settlements and density of population on the Deh Luran Plain, as well as the abandonment of more and more of the larger-scale water-control and irrigation systems after A.D. 800, generally follow the patterns found by Adams in Upper Khuzistan and Diyala. As yet the Deh Luran survey data have shown no evidence for the periodic upswings in canal building and settlement found by Adams (1962:120, 1965:84-111) for Khuzistan and the Diyala region. The Deh Luran survey did not recover any ceramic materials dating later than about A.D. 1250. Although no signs of warfare were noted, it is interesting that this date closely corresponds with the Mongol invasion of Baghdad in A.D. 1258. While Deh Luran itself may not have suffered the scourge of the Mongols, the threat of invasion may have sufficed to cause the nearly complete abandonment of this waning region.

## Sociopolitical Relationships

I would now like to elaborate upon the contention I made above concerning the directed engineering of Deh Luran water-control and irrigation systems. To submit this proposal or hypothesis in a more complete form will, I feel, better present the processes potentially involved in sociocultural change during the Sassanian Period. In turn, the implications suggested may provide additional information for the study and more complete understanding of the overall Sassanian sociopolitical organization.

I propose that population growth, the introduction of technologically advanced methods of water control and irrigation, and the sociopolitical developments reflected in the changing settlement and site (community) patterns from about A.D. 226 to 800 on the Deh Luran Plain are a direct result of processes of planned expansion and growth promoted by the Sassanian government.

The relatively little documentation available in translation indicates that the Sassanians undertook an intensive, well-planned expansion program to build the economy and population, and thereby the power and importance of their empire. These goals were apparently achieved with some success primarily through the efforts of the Sassanian kings Kavadh (A.D. 488 to 531) and Chosroes I Anosharwan (A.D. 531 to 579), who initiated the repair, planning, and construction of large-scale public works such as roads, bridges, and irrigation systems; the rebuilding and resettlement of entire communities; the incorporation of large numbers of war prisoners within the empire; and the encouragement of population growth by providing incentives for marriage and childbearing (Adams 1965:69-71; Rawlinson 1885:484-85, 488).

I suggest that, as part of this expansion, the Sassanians focused their attention on the Deh Luran Plain. This program could have been carried out for political and/or economic reasons to provide a location for the settlement of a population from outside the borders of the empire, or to relocate part of the rapidly expanding population from densely occupied areas such as Upper Khuzistan and the Diyala (Adams 1965:69-71). An alternative purpose for the program might have been to use Deh Luran as a "breadbasket" area, that is, to provide additional land for the production of badly needed foodstuffs for the rapidly expanding empire (Rawlinson 1885:488). In either case, or perhaps a combination of both, it is my opinion that trained Sassanian engineers made possible such an expansion on the Deh Luran Plain by adapting existing water-control and irrigation techniques to the particular environmental regime.[3]

An attempt to test these two explanations has indicated the first alternative, the location of new or surplus population on the Deh Luran Plain, to be the more tenable in light of data from the piedmont dry-farming homesteads and the alluvial plain irrigation-farming homesteads or hamlets. The sites DL-194, DL-274, and DL-275 as examples of these two types of farming units, and the ethnographic population-land use correlations presented by Adams (1965:23-25) were used to obtain the following results. Given a required ratio of 1.4 hectares of cultivable land per person, the household population size ranged from slightly greater to nearly two and half times the size expected for the area of cultivable land.

This result strongly suggests that the inhabitants of the plain during the Sassanian and Early Islamic periods were, *at best*, barely able to produce enough foodstuffs to maintain themselves. The surplus of grains needed to verify the second alternative mentioned above was far from present. However, the figures may be brought into somewhat better alignment, and the balance tipped to the side of a small surplus production, if two factors are considered.

A) Since the piedmont sites are without doubt based on dry farming, we may reduce the required amount of cultivable land per person to about half; i.e., about 0.7 hectare per person. This estimate may be made because the original figure of 1.4 hectares per person was calculated for a system in which there were equal parts of cultivated and fallow lands each year. The rationale for this adjustment is based on the fact that dry-farming land has periods of fallowing within each year and often does not require the relatively long periods of additional fallowing implied in Adams' 50-50 ratio. Further, the techniques of water control employed on the piedmont result in annual soil rejuvenation from small amounts of alluvium washed into the fields by runoff. Of course, yet another important problem to be reconciled is the difference in crop yield and the predictability of that yield, when comparing the irrigation techniques, upon which Adams' ratio is based, with dry farming.

B) While the ratio of 1.4 hectares of cultivable land per person may be fairly accurate for the alluvial plain irrigation-farming homesteads or hamlets, I feel the habitation area estimates are not as accurate as those for the piedmont sites. In the case of DL-274 and DL-275 (alluvial plain sites), the nature of the house mounds, and the poorly preserved state of the stone foundation generally surrounding the mounds, did not permit the determination of the exact size of the structures and the number of rooms. Having had to base my estimates on the total area of the house mounds and all of the surrounding debris, I would judge that I exaggerated the site size and thereby calculated too great a population for each of the homesteads or hamlets.

Since the marginality (or provinciality) of the Deh Luran Plain plays an important role in the formulation of my hypothesis, it seems appropriate that the reader understand the foundation for this contention. Basic to this discussion is the premise that sociocultural changes, including the introduction of material cultural items of "foreign" manufacture, usually take place more slowly outside nuclear centers of settlement, and economic and political activity, unless marginal areas are immediately or "personally" involved in the mechanisms of change. Evidence for such a time-lag in Deh Luran is seen in the following: (1) There are some indications that the transition from Parthian to Sassanian took place later in Deh Luran than elsewhere (i.e., post-A.D. 226). This observation is based solely on the ceramic data discussed below. (2) There was an apparent continuity of the Sassanian site occupation and settlement pattern until about A.D. 800. Sociopolitical and economic breakdown of the Sassanian Empire would have occurred first in nuclear centers because those were the first concentrated upon by the Islamic invaders. (3) The apparent delay in the introduction or acceptance of major technological innovations crucial to the subsistence activities of the Deh Luran inhabitants indicates marginality. I am, of course, referring here to the water-control and irrigation systems. (4) In the preliminary analysis of the Deh Luran ceramics, I received the distinct impression that changes in technology and artistic expression were slow to take place and were characterized by long periods of transition. Although possibly affected by my inexperience in the analysis of ceramics from these periods, I found it extremely difficult to determine any clear-cut divisions in the ceramic traditions or assemblages of ceramic types between the Parthian to Sassanian and the Sassanian to Islamic periods. The earlier ceramic tradition persisted while selected elements of the later ceramic tradition were slowly incorporated, and then, eventually, replaced their earlier counterparts. The marginality implied by the slow changes in the ceramic traditions is perhaps emphasized by the almost complete absence of luxury trade ceramics. This marginality is especially evident for the Islamic Period when trade wares from Samarra, Baghdad, and Samarkand are represented by only a handful of sherds.

If the marginality of Deh Luran can be accepted, we are provided with an interesting aspect of the Sassanian sociopolitical system. Several scholars, including Adams (1965:73, 82-83), state that a continuous process of sociopolitical integration or "centralization" of power has resulted or played a major role in the changes toward urbanization and increased population density. Yet, if my hypothesis is correct, and the various technological and sociopolitical conditions were at least several decades behind those in central and important regions of the Sassanian empire, then I submit there was a degree of sociopolitical autonomy, at least within certain segments of the empire, that has not yet been fully recognized or appreciated.

The very mention of "centralization" brings to mind the omnipresent problem of whether the term implies degrees of (or the presence of) sociopolitical autonomy at the lower levels, within the smaller segments comprising the system or empire. Our studies of sociopolitical processes are severely hampered due to the vagueness in the use of such terms in the current literature. Most assuredly, what I would call "large-scale" irrigation systems on the Deh Luran Plain are of a completely different magnitude than those larger networks so designated by Adams for the Diyala and Upper Khuzistan regions. I submit that a symposium to discuss and resolve such terminological problems is urgently needed. Even if agreement cannot be reached on single all-inclusive definitions, we would at least better understand the parameters connoted by their use.

Specifically because Deh Luran is now, and for most of its long history of occupation probably was, an out-of-the-way, marginal, or provincial region, characterized by a somewhat less-than-ideal environmental setting, it may provide extremely interesting and valuable information on the processes of acculturation involved in an early case of directed technological and sociocultural change. In addition, further detailed investigation in this region will provide data on certain aspects of the technological, economic, and sociopolitical systems of the Sassanian and Early Islamic periods not as readily accessible through the study of large nuclear centers. Of special importance will be the consideration of processes involved in the economic and social expansion of a large and developed political unit – the Sassanian empire. I plan future field work toward this goal as part of the ongoing archaeological study of Deh Luran.

## NOTES

[1]This brief and somewhat oversimplified presentation will be expanded in detail in the final report of the entire multidisciplinary project that took place in 1968-69.

[2]There is, however, another possible explanation for this apparent "marginality." It is possible that the growth pattern actually began earlier in Deh Luran than in the Diyala and Khuzistan regions. I have yet to assess fully this possibility since I have not completed my analyses of the Elamite occupations.

[3]The probable use of captive Roman legionnaires as laborers in the construction of irrigation and water-control systems during the Sassanian Period has been noted by Van Roggen (1905:168) and Adams (1962:116; 1965:69, 82-83). I am presently undertaking a comparative study, based on historical documentation and form-function analysis of Roman features comparable to those found on the Deh Luran Plain, which will assess the possible influence of Roman engineers on the development of these systems.

## REFERENCES

ADAMS, Robert McC.

1962 Agriculture and Urban Life in Early Southwestern Iran. *Science* 136:109-22.

1965 *Land Behind Baghdad: A History of Settlement on the Diyala Plains.* Chicago: Univ. of Chicago Press.

1970 Tell Abu Sarifa: A Sassanian-Islamic Ceramic Sequence from South-Central Iraq. *Ars Orientalis* 8:87-119.

AVITSUR, S.

1960 On the History of the Exploitation of Water Power in Eretz-Israel. *Israel Exploration Journal* 10:37-45.

COLT, H. Dunscombe, (Editor)

1962 *Excavations at Nessana (Auja Hafir, Palestine).* Vol. 1, pp. 1-24, 211-69. London: British School of Archaeology in Jerusalem.

CRESSEY, George B.

1958 Qanats, Karez, and Foggaras. *Geographical Review* 48:27-44.

ENGLISH, Paul Ward
1968 The Origin and Spread of Qanats in the Old World. *Proceedings of the American Philosophical Society* 112:170-81.

EVENARI, Michael and Dov Koller
1956 Ancient Masters of the Desert. *Scientific American* 194:39-45.

EVENARI, Michael, L. Shanan, N. Tadmor, and Y. Aharoni
1961 Ancient Agriculture in the Negev. *Science* 133:979-96.

FORBES, R. J.
1955 *Studies in Ancient Technology*, Vol. I. Leiden: E. J. Brill.

GREENE, Herbert
1966 Irrigation in Arid Lands. In *Arid Lands, A Geographical Appraisal,* ed. E. H. Hills, pp. 255-71. London: Methuen, and Paris: UNESCO.

HATT, Robert T.
1959 The Mammals of Iraq. *Miscellaneous Publications, Museum of Zoology, Univ. of Michigan,* No. 106. Ann Arbor.

HOLE, Frank, Kent V. Flannery, and James A. Neely
1969 Prehistory and Human Ecology of the Deh Luran Plain: An Early Village Sequence from Khuzistan, Iran. *Memoirs of the Museum of Anthropology, Univ. of Michigan,* No. 1. Ann Arbor.

KEDAR, Yehuda
1957*a* Water and Soil from the Desert: Some Ancient Agricultural Achievements in the Central Negev. *The Geographical Journal* 123:179-87
1957*b* Ancient Agriculture at Shiutah in the Negev. *Israel Exploration Journal* 7:178-89.

KIRKBY, Michael, and Anne V. T. Kirkby
1969 Provisional Report on Geomorphology and Land Use in Deh Luran and Upper Khuzistan. In *Preliminary Reports of the Rice Univ. Project in Iran 1968-69,* ed. Frank Hole, pp. 1-8. Houston: Rice Univ.

NEELY, James A.
1969 Preliminary Report on the Archaeological Survey of the Deh Luran Region, 1968-69. In *Preliminary Reports of the Rice Univ. Project in Iran 1968-69* ed. by Frank Hole, pp. 9-24. Houston: Rice Univ.

RAWLINSON, George
1885 *The Seven Great Monarchies of the Ancient Eastern World: The History, Geography, and Antiquities of Chaldaea, Assyria, Babylon, Media, Persia, Parthia, and Sassanian, or New Persian Empire.* Vol. 3, The Sixth and Seventh Monarchies. John B. Alden: New York.

U.S. Soil Conservation Service, Department of Agriculture
1959 Irrigation on Western Farms. *Agricultural Information Bulletin* No. 199.

VAN ROGGEN, D. L. Graadt
1905 Notice sur les Anciens Travaux Hydrauliques en Susiane. *Mémoires de la Délégation Archéologique en Perse,* Deuxième Série, VII:166-208. Paris.

WULFF, H. E.
1968 The Qanats of Iran. *Scientific American* 218:94-105.

CHAPTER 4

# IRRIGATION AND SOCIETY
## The Iranian Plateau

Brian Spooner
*Department of Anthropology, University of Pennsylvania*

Every anthropologist needs to start out by considering just how much of the culture with which he is faced can most readily be understood as a direct adaptation to the environmental context including that part of the context which is man-made. [Leach 1961:306]

Although it is generally taken for granted that "irrigation societies" constitute an obviously valid category for cross-cultural study and sociological and historical generalization, the results so far have been notably meager. For this reason alone it might prove useful to trace the categorization back through the literature, to see how it arose, and to consider whether a different or at least a modified categorization might facilitate generalization.

As a category for anthropological study, the term "irrigation civilizations" appears to owe its currency to the symposium edited by Julian Steward under the title *Irrigation Civilizations: A Comparative Study, a Symposium on Method and Result in Cross-cultural Regularities* (1955). This symposium was organized at the 1953 Annual Meeting of the American Anthropological Association at Tucson, Arizona with the explicit aim of testing the "basic methodology" of studies in cultural evolution (Steward 1955:1). The interest which generated the proposal for such a symposium is traced to Leslie White's neo-evolutionary theories and to V. Gordon Childe's efforts to "interpret Old World archaeology according to a general evolutionary approach." The particular theme "irrigation civilizations" was chosen because of "the seeming parallels in the development of the early irrigation civilizations of Mesoamerica, Peru, China, and the Near East" (Steward 1955:1-2):

The hypothesis that these civilizations had developed through similar periods for fundamentally the same reasons provided the hypothesis to be tested. This hypothesis is evolutionary, but it arises from a special conception of cultural development known as multilinear evolution. . . .

The concept of multilinear evolution was provisionally formulated in "Culture and Process" prepared for the 1952 symposium of the Wenner-Gren Foundation (Steward 1953). The germ of this concept, however, was stated nearly twenty years earlier in "The Social and Economic Basis of Primitive Bands" (Steward 1936). This approach received tremendous stimulus when Karl Wittfogel formulated the cross-culturally recurrent characteristics of the type that John Stuart Mill had called the "Orient State." After devoting many years to detailed analysis of Chinese culture history and thoroughly aware of the importance of water control in several key world areas, Wittfogel (1938) described this type of state as an Hydraulic Society or Hydraulic State.

In 1949, I undertook to extend Wittfogel's formulation by exploring the possibility that the hydraulic or irrigation societies began a parallel evolution with the first use of domesticated plants and that the development of local communities, technology, and even intellectual, aesthetic and religious achievements as well as the economic and political patterns ran similar courses.

At first sight the subject fits the proposal perfectly, and of course the resulting symposium was an excellent arena for the extension of Wittfogel's theories. However, quite apart from the merits or demerits of the Wittfogelian hypothesis, neither "irrigation society" nor "irrigation civilization" was explicitly defined, and none of the contributors to the symposium suggested any logically consistent reasoning for the definition of such a category for anthropological inquiry – beyond the "seeming parallels" of development. In fact the symposium was concerned exclusively with historical populations that depended economically upon types and scales of irrigation engineering requiring a relatively high level of investment; examples of populations depending on irrigation technologies that require relatively low levels of investment were not considered.

Since the publication of Steward's symposium, there has been a considerable increase in both the quantity and the quality of available data (ethnographic and archaeological) on populations practicing irrigation, and Steward (1960:323) himself has revised his position. However, very little, if any, advance has been made in the theoretical or conceptual framework in which they have been published, and (to the best of my knowledge) the literature still contains no definition or justification of the categorization. This essay attempts to clarify these issues. The approach is ecological, and the method is first, to suggest a number of analytical distinctions which should be made in the anthropological study of populations dependent upon irrigation, and second, to illustrate their usefulness in brief accounts of two villages on the Iranian plateau. In so doing I hope to make some small contribution in general terms to our understanding of the systemic relationships between traditional technologies of irrigation and the organization and structure of the societies that have depended upon them.

## ON THE STUDY OF "IRRIGATION SOCIETIES"

Irrigation implies a three-way relationship – men-land-water – in a temporal dimension. It therefore represents excellent subject matter for an anthropological study under the heading of "ecology and social structure" (cf. Forde 1971). For the present purpose we may define it as a problem: how much water must be led on to each of how many parcels of land by how many men how often. It is immediately obvious that we have four problems in one. The consideration of each of these problems in relation to the others should allow the formulation of basic criteria for a typology of irrigation systems. Typology is not, of course, an end in itself. It is, however, at least implicit in the first stage of generalization. In this context, it is well to make it explicit, since it is the underlying thesis of this essay that the anthropological literature on irrigation has suffered from implicit and inadequately thought out typology.

The first distinction that must be made is between *irrigation* and *inundation*. Inundation is the technique of maintaining a certain depth of slowly but continuously moving water as in wet rice cultivation in southeast Asia, where the crop acquires most of its nutrients not from the soil but from the water (see Geertz 1963:28-37). In irrigation, the nutrients come primarily from the soil, and the crop is watered at certain intervals to make up for deficiencies in the rainfall. This essay is concerned with irrigation rather than inundation. However, in certain cases inundation systems will be cited as examples of technologies which could be used for inundation or irrigation.

The next distinction which must be made is between three subsystems which are present in any system of irrigation. These are the distribution of water flow over the land parcels, the distribution of water rights among the population and investment. The first of these is temporal, the second social, and the third economic. All three subsystems are of course closely interrelated, but their relative significance in the relationship between irrigation and society varies according to the total technological context.

The essay by Netting (this volume) and the example of Deh Salm later in this essay adequately show that the maintenance of an efficient temporal distribution system not only does not require explicit or centralized control, but no one person need be aware of the full details of the system. In the case of Deh Salm, this may be explained by reference to the systems of inheritance and bridewealth, the transferal of property at marriage. The total number of shares is immediately divided into blocks which are controlled within groups of families who are more closely related internally than they are with other similar groups. I found that any informant had a general knowledge of which social or political grouping of families controlled the water flow at any given part of the cycle.[1] This knowledge was sufficient for the smooth working of the system; the method of division within the grouping was irrelevant to people outside it. Moreover, keeping the distribution system working – and the water flowing – never required more than one man to work at a time. No cooperation of individuals, let alone groups, was needed. Even cleaning the main channels could be accomplished by individuals. Therefore, no organization was required to facilitate the cooperation. During a three-month field season, I did not encounter any case of one individual infringing upon the rights of another by keeping the water flow beyond the time limit of his share. I was told it never happened, and I deduce that it is in every individual's interests to keep the system working smoothly. Thus, rights to a particular feeder channel or small water source could be shared by two spatial communities (whether or not they are socially or politically separate villages, cf. Netting, this volume, and below, the case of Nayband), without generating any explicit authority system for central control. However, in these and similar cases of complex temporal distribution (cf. Lambton 1953, Chap. 10; Leach 1961), there is always the ultimate sanction, at least at the time when the studies were carried out, of appeal to a higher authority – police, law, government. This authority, however, is not directly concerned in the maintenance of the irrigation system, and it is difficult to find examples of its employment. It might perhaps be argued that the Sonjo provide a unique case of local systems of centralized control, where without them there would have been no ultimate sanction (Gray 1963). However, even in this case, it is not clear whether the social structure developed primarily in response to the irrigation system or to the external Masai threat (cf. Millon 1962:62). In any case, it seems possible to propose as a minimum generalization that since any disturbance of the temporal distribution system affects all shareholders adversely, the normal premium on social order is increased.

With regard to the social distribution of water rights among the population, the situation is similar. Once again the data do not support the hypothesis that centralized policing or authority is necessarily generated. This aspect of the effects of irrigation technology on the organization of society must be sought in another direction – the process of individuation of rights in relation to resources. In recent years there has been a tendency to discuss such evolutionary processes in the context of population growth (e.g. Boserup 1965, Flannery 1969, Spooner 1973). In this context the most apposite statement so far has been made by Dumond (Spooner 1973:300-1):

> . . .pressure on basic resources tends to tighten concepts of tenure, so that in place of free use of plentiful land, the rights of use are ultimately vested in a corporate group that itself has rights of reversion when usable land is not employed. If this is true, it seems reasonable to think that continued pressure will further increase both the demand for land and the temptation to control it privately (see Boserup 1965:86-87). *The temptation should be especially great when land is irrigated.* Irrigation can be seen as a population-density-related, labor-produced capital improvement, that adds to the value of the land. Its importance is that it sharply defines an area of high land value. It is important in its influence on matters of land tenure, but not because of the managerial requirements of the irrigation system itself. . . . (Emphasis added)

On the basis of this brief discussion of the temporal and social subsystems, it may be argued that the features of irrigation technology which are most significant in the organization and structure of irrigation-based societies derive from the economic subsystem, or the investment factor, and it is on this factor that the remainder of this essay is focused.

The decision to irrigate, like the decision to cultivate, represents a decision to increase the productivity of a piece of land. It invariably involves some degree of capital improvement or investment. These two factors of the economic subsystem – increase in productivity and creation of investment – either in combination or severally, lead to certain predictable social consequences, which are more tangible than the order and individuation predicated for the temporal and social subsystems. We might expect that the greater the investment the greater the emphasis on title and control, and, consequently, the greater the degree of individuation and stratification in the particular society. Similarly, different types of investment might be expected to generate different systems of ownership and control.

For instance, in the study of the traditional systems on the Iranian plateau, it is possible to argue that an increase in productivity and investment brings about an increase in the value of land, and creates problems of differential access and title. Title tends to be individuated. The greater the investment required to ensure a continuous flow of water for irrigation, the more valuable will be title to water compared to title to land. There are examples in which the available cultivable land is as restricted, and even requires as much investment, as the water flow, and (as we might expect) title to land automatically includes title to water (for example, in the case of Nayband, below). Where the available land is less restricted than the water, title to land may be separate from title to water, and in extreme cases where the available land is far greater than the available water, title to land may be included in title to water. In small irrigation systems, especially where the engineering problem is to start and keep the water flowing (e.g., qanat, see English 1968 and Neely, this volume) and not simply to provide channels for water (as in, for example, Mesopotamia and the really "hydraulic" systems which attracted Wittfogel's avowedly "geohistorical" rather than ecological interest), the tendency is for water to become the object of individual title. In such situations, the effects on society of the process of individuation of title to resources is intensified. Generally, if a man owns land by a river, he gets water anyway, but if he owns land by a qanat, he does not necessarily get any water from it. Qanats are privately owned, but rivers are not. The formulation of this situation in Islamic law is both logical and instructive (Lambton 1953:210-11):

According to Islamic theory water is divided into three categories: that coming from (1) rivers, (2) wells, and (3) springs. Rivers are further subdivided. First are great rivers, such as the Tigris, the Euphrates, and other large rivers, the water of which suffices for all the needs of cultivation and from which anyone can lead off a channel to irrigate his land. These are owned in common by all Muslims. Secondly, there are lesser rivers, subdivided again into (1) those the water of which is sufficient to be led off without dams to irrigate the land situated along the river banks, and from which canals to irrigate other lands can only be led off if such action does not prejudice the position of the lands situated along the banks, and (2) those in which barrages have to be made, in which case lands situated higher up the river have a prior right to those situated lower down. The amount of water which can be taken off depends upon circumstances, local needs, and custom. Thirdly, there are canals dug to bring water to dead lands. These belong to those who dug them.

Qanats are privately owned because they are constructed by private individuals. By similar reasoning, rivers may not be privately owned, and since they cannot be private property, individuals do not invest in improving their efficacy for irrigation, even if it is within their means. The improvement is done only by states, which by definition own all land and water within their territory that is not privately owned, as well as holding reversionary rights over privately owned land and water. The only exception to this rule is the small river or stream in which the amount of water flow produced by the investment does not greatly exceed the requirements of one man's cultivation. Streams may, however, be used to produce "hydraulic" systems – as, for example, in Ceylon (Leach 1961:16) where the classical Sinhalese kingdom, with its capital at Anuradhapura, was a striking and characteristic example of what Wittfogel has called "hydraulic civilization" (Wittfogel 1957; cf. Leach 1959). Here a region of poor natural fertility was made to support a large and flourishing population by resort to irrigation engineering. In this case, however, the engineering problem (Leach 1961:17-18) was different from that in Iran:

The modern irrigation works of Nuvarakalaviya, like their ancient predecessors, fall into two distinct categories. There are the small reservoirs (tanks) associated with individual villages and the very much larger central reservoirs and feeder canals which now, as formerly, are under the control of the central government. . . . Pul Eliya is not today connected with any central irrigation system and, so far as can be judged, it never has been so connected in the past.

Of the smaller reservoirs – the village tanks – there are today several thousand in actual use. Almost all are of ancient origin, but only a few have been in continuous use over the centuries. The great majority have been abandoned at various times and then restored again.

In this economy the basic valuable is scarce water rather than scarce land. . . .

A village tank is created by damming up a natural stream and building a long earthwork wall to hold the water up behind it. The resulting reservoir (when full) is usually about seven feet deep immediately behind the earthwork ('bund'). Very roughly, the full tank covers much the same area of ground as the land below it which it is capable of irrigating. . . .

The village of Pul Eliya today has a main tank of about 140 acres.

In this study of Pul Eliya, Leach (1961:1, 20) exaggerated the role of the land and water tenure system in social organization in such a way that he ensured that it should never again be ignored. It was not his aim to contribute to the type of discussion which is the theme of this volume. Nevertheless, he includes the information that at least the older part of the system is not really centralized (cf. Millon 1962:64) even though it is highly complex. The fact that it is complex (in terms of the relationship between division of land, division of water, and distribution of rights to both among the population) is particularly significant here, because it is at least partly for this reason that despite the smallness of scale (the total population was only 146, and the irrigated acreage 135), repairs to the tank had to be government-assisted. It is argued below, in the case of Deh Salm, that small complex systems tend to have investment problems which require or invite political or economic interest from outside the community.

It follows from these considerations that when discussing types of investment, a third distinction must be made between the engineering of 1) *conveyance*, by channelling water which is already naturally flowing on the surface, 2) *storage*, by means of dams or tanks, of water which although already on the surface is not always available when it is needed, and 3) *bringing to the surface* groundwater by qanat or wells. Many systems must cope with more than one of these problems. Nevertheless, one generally tends to be more characteristic of a system than the others.

At this stage it may be useful to draw up a rough typology of irrigation systems based on source, in order to bring out the relationship between the source, the primary engineering problem, and the other criteria which emerge from the discussion:

I. Permanent flow surface sources
   A. Large permanent rivers
      1. Basin irrigation, e.g. the Nile[2] (cf. Kees 1961:52-60; Hamdan 1961).
      2. Canal irrigation – perhaps the most common of all, e.g. Mesopotamia (cf. Jacobsen and Adams 1958, Adams 1965, Fernea 1970), the Indus (Leshnik 1972), China (cf. Te-K'un III:xxx-xxxi, Chi 1936), Mesoamerica (Steward 1955; Price 1971), Peru (Steward 1955), Khwarazmia (Frumkin 1970, Chapter 5), Sistan (cf. Le Strange 1905, Chapter 24).
   B. Small rivers and streams, e.g. the Sonjo (Gray 1963).
   C. Springs, e.g. Nayband (see below).
II. Intermittent or extremely variable-flow surface sources
   A. Sheet run off or wadi flooding, e.g. Hadhramaut (Hartley 1961, Bujra 1970), Baluchistan (Raikes 1965).
III. Ground Water
   A. Qanat (kariz, foggara), e.g. Iranian plateau (cf. English 1968), Sahara (Briggs 1960: 10-12), cf. Deh Salm, below.
   B. Wells, etc. which require lifting techniques such as the shaduf (cf. Drower 1954).

The fourth distinction, which may be introduced now, is that of scale. This is of course not a new distinction, and it has been implicit already in this discussion. However, the distinctions of scale that have been made in the literature are unsatisfactory because they are either arbitrary or have been derived from the theory they were designed to prove, rather than from the technology. Wittfogel (1957:3) made the distinction between a "farming economy that involves small-scale irrigation" which he called "hydroagriculture" and "one that involves large-scale and government-managed works of irrigation and flood control" which he designated "hydraulic agriculture." The distinction is of course not without validity, but requires more careful definition based on ecological rather than bureaucratic premises. Wittfogel (1957:17-18) is not concerned with hydroagriculture:

> Irrigation farming always requires more physical effort than rainfall farming performed under comparable conditions. But it requires radical social and political adjustments only in a special geohistorical setting. Strictly local tasks of digging, damming, and water distribution can be performed by a single husbandman, a single family or a small group of neighbors, and in this case no far-reaching organizational steps are necessary. Hydroagriculture, farming based on small-scale irrigation, increases the food supply, but it does not involve the patterns of organization and social control that characterize hydraulic agriculture.

Though the "hydro-" terminology has found only limited acceptance, the "large-scale/small-scale" distinction is common in the literature, and has been modified only slightly, for example, by Millon (1962) who also distinguishes single village systems from multi-community systems. I wish to introduce a further modification, or rather set of modifications, for in the context of the relationship between irrigation and society, the criterion of scale works differently depending upon whether we are considering the economic, temporal, or social distribution subsystems of the total irrigation system. The latter two have been treated briefly above.

In the consideration of the economic subsystem, there is a more obvious case to be made for the type of scale distinction that has generally been used. But Wittfogel's equation of "large scale" with "government-managed" is unnecessary and misleading. I suggest that if we start from ecological premises, since we have seen that centralized authority is not necessarily generated by the water distribution systems, we have no grounds for positing that it should be generated by the investment subsystem. We should look instead at the degree of cooperation that is required to keep any given subsystem going, and whether or not, when we find cooperation, we also find organization of cooperation. In the temporal and social distribution subsystems, positive or organized cooperation is not required, whereas in the economic subsystem it often is required.

So long as the investment required to maintain the system is within the means of the individual operators, it may remain a village system. However, when a differentiation of wealth arises in the population to the degree where a single operator is able to take on a broader investment role in the community, a qualitative sociological change may take place. Assuming that the incentives are there, one or more men may, through investment, acquire controlling influence over productivity and labor beyond their subsistence requirements. Thus, an economic elite is formed which, though it may not lead to explicit differentiation of political status and centralization of power, must affect the political situation.

Such differentiation of individuals may take place without affecting the status of the system as a single village irrigation system, operated, maintained, and controlled from within. However, in many such cases the differentiation did not develop within the system, but was imposed upon it from without. This exogenous imposition of a "class" system can take place by political or economic means – by conquest or by economic imperialism. Such cases are typically based on the control and diversion of larger rivers. However, it is necessary to note first, that not all large rivers are exploited exclusively by large scale systems (e.g., basin irrigation on the Nile, in Kees 1961:52-61), and second, in qanat systems it is possible for one man to acquire a controlling interest in the qanat of several villages – a situation which was common in Persia before Land Reform. It should also be noted that this latter case is essentially one of economic, not political, centralization. It requires a great degree of financial investment, but it does not require a high degree of organization of men and materials, for qanat construction takes much time, but cannot utilize large numbers of men. Of course, ownership of the water supply of even one village – let alone of many villages – cannot be without political significance. However, this political significance is of the order discussed above in relation to the social distribution subsystem. It is indirect, not direct political control as in the Wittfogelian scheme.

Large scale systems, therefore, tend to differ from smaller systems in the degree of centralized organization – of money, men, or materials – which they require for maintenance and expansion. Therefore, from the sociological point of view, rather than distinguishing according to size (small scale and large scale or single village and multi-village systems), it would seem more fruitful to make an economic distinction based on complexity. Systems in which the individual operator can control and maintain his own part, whether or not he knows the details of the total system, I call "simple systems." Whereas, "complex systems" require engineering and maintenance beyond the ability of individuals or groups involved in actual cultivation.

To return now to the typology given above: in type I, permanent flow surface source, the engineering problem consists of conveying the water by gravity flow in a controlled manner from the source to the fields. In type I.A.2, canal irrigation involves the digging, maintenance, and cleaning of canals or channels (cf. Fernea 1970:57-162). It is in this category that we find the greatest range of scale, but in the terms proposed above, this category comprehends both simple and complex systems. In type II, intermittent or extremely variable-flow surface sources, the engineering problem lies in the storage and control of water which would otherwise not be available at the desired season or in the desired quantity. In type III, groundwater, the problem is to bring the water to the surface. In terms of absolute scale, systems which fall within types II and III are generally small, and would be accounted "small scale" according to the traditional large scale/small scale criterion. However, they may be either simple or complex. Qanat systems do not generate centralized authority on any scale, and have therefore been distinguished conceptually from the hydraulic systems which would fall into type I.A. This distinction is, however, quantitative, not qualitative, and from the anthropological point of view, it is primarily the qualitative distinctions that are significant, since any useful model of the relationship between irrigation technologies and other aspects of culture and society must be constructed in generative terms, and the basic typology or scheme of definitions must be subservient to the generative model.

As soon as the necessary degree of organization of labor, materials, or money is beyond the means of the individual operator, some degree of political or economic centralization becomes adaptive. However, centralization does not always develop, and it is in fact very difficult to show that any centralized system has actually evolved in this way. It is much easier to demonstrate the reverse, for example, that centralized political systems have been able to initiate and maintain complex irrigation systems because of their ability to organize investment on a large scale.

If the argument so far has been valid, and the analytical distinction between simple and complex irrigation systems is as far as we can go in generating social organization from ecological considerations in societies that depend on irrigation, then the situation in societies with simple irrigation technology but centralized political or economic systems must be explained by factors other than irrigation. Further, if there is nothing inherent in irrigation per se that has predictable effects in other aspects of society and culture but only in the organization of labor and materials involved in certain types of irrigation, then there could logically be other technological systems which are comparable in their social effects. It seems likely, however, that no other such technologies evolved before the modern industrial age.

## TWO ETHNOGRAPHIC SKETCHES

These sketches illustrate the application of the simple/complex criterion in the analysis of two small oasis villages on the Iranian plateau in eastern Iran.[1]

## Deh Salm

The village of Deh Salm lies in an extensive arid plain some 100 kilometers west of Neh, which is the southernmost administrative center (*bakhsh*) of the province of Khorasan in eastern Iran (see Neely, this volume, Fig. 1). It has a population of forty families. There is no other settlement in the plain, and only a few small villages in the mountains to the north and northeast. To the east, between Deh Salm and Neh, over a range of mountains and a difficult pass, lies a plain where pastoralists formerly practiced small-scale grain cultivation with the excess water from wells (Chah Dashi) used to water their animals. Chah Dashi has recently been developed for grain growing by means of diesel pumps. In the remaining arc from the southeast round to the northwest, there is no further settlement until the other side of the desert, a minimum straight distance of 150 kilometers. To the south, the territory has traditionally been under the control of Baluch tribes of nomadic pastoralists. Deh Salm therefore represents the southwestern outpost of settled agriculture in eastern Iran. Its geographical isolation has been an important factor in its history. However – and this is the primary point that the following ethnographic and historical summary serves – it represents a type of agricultural settlement which cannot exist for long in economic and political isolation.

There are no rainfall records for the area; however, I estimated it to be well below 100 millimeters per year, with great variation from year to year and highest probability in spring and autumn. For agricultural purposes – the main economic interest of the population as a whole – rain is simply a bonus if it should fall at an agriculturally useful time. Whenever it falls, however, it brings peripheral and indirect benefits by transforming the surrounding countryside into good pasture (on the rather low local standards), especially for goats and camels.

Rainfall does provide the main source of drinking water. Runoff is channelled into small domed tanks (*hauz*; elsewhere called *āb anbār*; for details of construction, see Siroux 1959), where after the sediment has settled it remains fresh and clear for many months. This water is referred to as "the water of God's mercy" (*āb-i rahmat*), and should be used only for drinking, not cooking, washing, or irrigation.

Essentially, Deh Salm depends upon a single water source – a qanat (see Chap. 3). The qanat enters the village from the north, and surfaces in a stone-lined pool which is used for general washing purposes by the men. From here it passes into another similar pool surrounded by a higher wall, for the use of the women. Beyond this pool, the qanat divides into two main channels which serve the two tongues of cultivated area (*keshmun*) and palm plantations which stretch out to the south. The short stretch of some fifteen yards from the men's pool, past the enclosed women's pool, to the division of the channels is the hub of the village. All changeovers between shares of water flow for irrigation are taken from here.

The most significant factor in the pattern of land distribution and use is the distance from the point where the channel divides (*sar-i ju*). The farther the water has to flow to the crop, the more water is lost through seepage and evaporation on the way. (However, no time of flow is lost, since the next man – even if his plot is adjacent – must start his flow also from *sar-i ju*.) Therefore, the land closest to the divisor (*sar-i ju*) is the most valuable, and may sell for as much as 5000 Rials per *faran* (approximately 600 square metres; 75 Rials = approximately $1.00; the price of a kilogram of wheat varies from 10 to 15 Rials). Land farthest away may sell for only a fifth of that price. Palms are grown throughout the cultivated area (*keshmun*) along the sides of channels and on land unsuitable for cultivation, but they are raised only on cultivable land on the cheaper stretches at the limits of the *keshmun*. The land has had to be dug down, in some cases several feet, in order to lower the surface to a level which is irrigable by gravity flow from *sar-i ju*. This leveling represents the only major investment entailed in the creation of fields. All that remains to be done is to divide the land into irrigable segments (*kal;* elsewhere in eastern Iran, *khid*) connected by feeder channels (*ju*). The ideal segment size is considered to be about half a *fārān*, i.e. approximately 30 x 10 meters, although many are smaller. Each segment is separated by a ridge of earth (*pal, bāza*), and when a segment is owned by different people, stones are laid along the top (*sang-i bāza*). A simple scratch plough is sometimes used by men who have large holdings, but digging with spades is considered preferable. A common type of dispute arises when a man ploughs or digs consistently a little too close to the ridge so that a little earth falls over it onto his neighbor's side. As a result, the ridge itself gradually moves over. It was noticeable that ridges generally tend to meander slightly. Digging is done to a depth of six inches or more and is extremely hard work, especially in summer temperatures of well over 100 degrees F. in the shade.

Land is fallowed every other year. The summer is the main time for digging because the land has to be prepared for the sowing of the autumn crop. The young men work together in informal teams of three or four, digging each other's land. Older or richer men, or women whose men are away, hire a digger at the cost of a day's food, including an evening meal with meat. A man starts early in the morning, works till nine or ten A.M., and returns for about two hours in the afternoon. An able man might finish half a *fārān* within the day.

Before digging, each segment is watered to soften it. After digging, it is left to take in the sun (*āftāb bokhorad*) for a month or so before being planted. The major crops are wheat and turnips. Several crops may be sown on one piece of land in the course of the year, starting with turnips in the autumn and finishing with melons in the summer. Then the land will be left fallow for a year. Cotton is grown on the same land several years running without the intervention of other crops, except that melons may be grown amongst it in the summer. Garden vegetables are not grown in the village except by a gendarme on borrowed land, with borrowed water.

There are an estimated 25,000 date palms in Deh Salm. If the count is not too far out, in an "average" year the palms should yield approximately 1,250,000 kilograms of dates of varying species and qualities. The best dates sell at 10 Rials per kilogram. The best trees are priced at 5,000 Rials, though an average price would be in the region of 3,000 Rials. Like tree crops generally, dates require very little labor for a high return. They must be artificially pollinated in the spring and picked in the late summer or early autumn according to species. Apart from these two operations, which require large amounts of labor for short periods, one man can tend literally thousands of palms. According to their situation, date palms may require occasional irrigation. Dead or excess branches and clusters of fruit should be removed, and during the ripening season care should be taken that the heavy clusters are properly supported. Because of the low labor requirement, palms with potentially large annual produce, however important they could be in a market-oriented economy, do not provide the principal occupational focus of the population. Similarly, they are marginal to the irrigation system.

Besides land, water, and date palms, the average family owns a few goats and perhaps a cow and a camel. Many more camels belonging to non-villagers (estimates range from one to three thousand) roam the surrounding plains. These find their own pasture, returning to the village for water at intervals from a day to several weeks according to the season and the state of the pasture. Occasionally their owners come from elsewhere and round them up to sell as meat.

Scarcely anything is known of the history of Deh Salm, but it is generally considered to date from at least the early Islamic period. The circumstances of the construction of the qanat are unknown. It is approximately three kilometers long, and it was necessary to use pottery hoops along at least part of its length to strengthen the roof of the channel. However, routine maintenance has not been carried out in it for some time since there are no professional qanat-diggers (*moqanni*) in the village, nor anyone with any special interest or expertise in qanat construction or maintenance. Throughout the period I spent in Deh Salm, I could elicit no data about routine maintenance, and I am confident that if ever anything had to be done to the qanat it would require help from outside the village, both for labor and financing.

The flow of water is distributed according to a twelve-and-a-half-day cycle. The cycle is divided into twenty-five periods (*meh*) of half a day each, from sunrise to sunset or sunset to sunrise, and each period is divided into twenty-five shares (*khābiya or khomma*, literally "pitcher"), making 625 shares in all. When evaluated according to the clock, each share is considered to last half an hour.

Record of the distribution of shares is said to have been kept in a book. However, if this is historically true and the book still exists, no one seems to know where, and it has not been seen for many years. In day-to-day distribution, the length of each share used to be measured in the conventional way by letting a small metal bowl with a hole in it sink in a larger bowl filled with water. Each clink of the smaller bowl hitting the bottom of the larger bowl signalled the end of a unit of water flow. However, this method is no longer practiced. Instead, the time is estimated according to the sun, and when there is any doubt, someone is sent to ask the time of the gendarmes, or of myself! Such vagueness is possible because in only very few cases does the ownership change at the end of only one share, and when it does, it is either rented to someone else who owns at least one share before or after it in the cycle, or the owner forms a "company" (*sherkat*) with those owning the shares on either side. In the latter case, a number of men agree to pool their shares of water and cooperate in the cultivation of their land. It might, therefore, be argued that the poor are obliged by the temporal subsystem to surrender part of their autonomy.

The use of a twelve-and-a-half-day cycle cancels out the inequalities which resulted from the varying lengths of the night and day during the year. Sunrise and sunset are always used as the points of reference from which the shares were estimated in terms of hours and half hours.

Time estimation is not the only element of vagueness in the operation of the system. No one owner – so far as I could tell – knew the total distribution system. Though disputes are a regular part of daily life in Deh Salm, they were never (while I was there) occasioned by differences over the distribution of water. However, while no one can recite the succession of owners through the total cycle, every member of the community measures time in terms of twelve-and-a-half-day cycles and associates different parts of the cultivated area (*keshmun*) and social groups in the village with different parts of the cycle.

The system has no overseer or referee. When a man considers that it is time for his share, he proceeds to the divisor (*sar-i ju*) and either diverts the flow from one channel to the other, or if that is not necessary, follows the flow to the point where the change has to be made. He walks along the channel in front of the water, clearing it out to improve the flow at points where animals have trodden or the sides have collapsed, until he arrives at the set of segments (*kal*) he wishes to irrigate. Depending on the distance of the segment from the division, it might take as much as an entire half hour share (*khomma*) to water the first segment, for if the channel is dry it absorbs a great deal of the water on the way. Once the channel is saturated, however, one share will be sufficient for at least two segments.

One share of water sells for 4,000 Rials, but it may be rented for a whole year for a fixed rate of 36 kilograms (12 *man*) of wheat. However, if land and water are rented together, the owner takes two-thirds of the crop. In many cases the distribution of water does not fit the distribution of land: an individual's land shares may be distributed in several different parts of the cultivated land, whereas his water shares may come one after the other. This inequality of distribution leads to a great deal of swapping of water between friends.

As might be expected, the actual situation is more complicated than this description would imply. The complications arise from the way title to the shares of land and water fit the pattern of operation. Approximately one-half of the water and land is held by members of the community of Deh Salm. Likewise, villagers own merely one-tenth of the date palms and a negligible number of the camels. The average family has only a little land, a little water, and a few palms, acts as a tenant for more, and owns a few sheep or goats, and perhaps a cow and a camel. Before Land Reform was begun in 1962, this situation would not have been unusual in Iran. However, Deh Salm was untouched by Land Reform because the resources which are owned outside the community are distributed among a large number of piecemeal owners – not held in one block by one or a few large absentee land-owning families. The picture is further complicated by the fact that several of the families of Deh Salm own small pieces of property in the mountains north and northeast of the village, and some of the poorer men migrate to find work in Chah Dashi and elsewhere in slack seasons. The way this unusual situation arose is of especial interest in the context of this volume.

In order to run smoothly, the Deh Salm irrigation system requires, besides a minimum annual maintenance, provision for organizing repairs after irregular disasters, as, for instance, when runoff engulfs one or more of the qanat well openings and fills in part of the qanat. In similar systems where the greater part of the qanat belongs to a single owner, such disasters present no great difficulties, but where shares in a qanat are fairly evenly distributed throughout a community, very few communities show a successful maintenance record. This problem has been perennial among small holders in Persian villages based on qanats, and was one of the main arguments used by landowners against Land Reform in the early sixties. This problem is particularly acute in eastern Persia, where there is less provision for corporate action in village communities than elsewhere on the Iranian plateau.

This lack has been compensated to some extent by the development of a form of political organization based on what I choose to call the "dynastic family" (Spooner 1969). In the manner suggested by Parsons (1964:161-2), groups of villages in eastern Persia tend to be politically integrated by a small elite oligarchical superstructure of families who maintain their status and identity by forming a network of marriage alliances with other families of similar status over a certain area, rather than marrying within the village which they largely control (see especially the case of Gunabad in Spooner 1965). The villages themselves are otherwise largely endogamous. Where there is a "core" village, larger than the rest, this village is invariably the seat of a leading dynastic family with minor lines or allied lines in subsidiary villages in outlying areas. Thus, during the last century, first Qā'en, and later Birjand, was the home of the major dynastic family in eastern Persia, while subsidiary, politically allied lines were dominant in Tabas, Gunabad, and Neh. The dynastic families claimed tribal origin, typically justified by a genealogy going back to their entry into the area in the seventeenth or the eighteenth century. The agricultural population which they dominated was not genealogically organized. Because of their dominant position, they gradually acquired title to large amounts of property distributed throughout the hinterland of their agricultural centers, but this title was not the original basis of their position. Their position was rather of a feudal nature, in that they exacted political allegiance and a portion of the agricultural produce from their respective areas in return for political patronage, military protection, and a certain amount of redistribution of wealth.

A subsidiary line of the family which dominated Neh (the largest center in Khorasan south of Birjand) lived in Deh Salm. As long as there was a khan, a dynastic head, resident in Deh Salm, his relationship with the dynastic family in Neh insured the defense of the village. He provided a focus within the community for leadership, and – more important – for the organization and investment of labor and materials in the maintenance of the qanat. But at some point in the latter half of the nineteenth century, the khan left Deh Salm.

The circumstances of the khan's move from Deh Salm are unclear. Historical events are difficult to fix chronologically, and information about them is difficult to elicit beyond the manner in which they affect the community. It is evident that the plain in which Deh Salm is situated once held a considerably larger and more prosperous population than it does now. The remains of three qanats are still visible within five kilometers of the village. There are signs of earlier cultivation and isolated ruins of relatively well-constructed buildings. According to oral tradition, an earlier and more prosperous settlement had been located in the southeastern part of the cultivated area (*keshmun*) of Deh Salm. Here and a hundred yards or so beyond, small broken potsherds are strewn. Several men spoke of finding beads when they dug in that part of the *keshmun*. The oldest villager claims to remember that a line of palms once stood at the end of one of the derelict qanats. It is possible, therefore, that the decline which led to the neglect of these investments was associated with the move of the khan away from Deh Salm, although, since there is no way of dating the ruins with any certainty, they may relate to a much earlier period. The khan's residence in Deh Salm is now in a state of ruin similar to that of the buildings out in the plain, but the dilapidation could have been accelerated by the practice of transporting mud from abandoned houses to the fields.

After the khan's departure, the villagers were left with no superstructure of authority. They were thus immediately deprived of their means of organization for defense, their primary structural relationship with the larger, protective settlement of Neh, and their means of organizing labor and materials for maintenance and repair of the qanat.

There were also other factors (the importance of which has been overlooked or underestimated) which developed at about the same time, and must have added to the plight of the village. One of these was the introduction of tea, which gradually became a staple element in the diet of the ordinary peasant throughout Iran. Since the peasants could not grow their own tea, they became, for the first time, dependent on an outside market source for a staple item of their diet. The second factor was the introduction of opium, which was soon used as a panacea. Since malaria was endemic and caused great hardship, particularly in the summer, addiction became commonplace. This situation seems to have led to a few of the more affluent families of the village acquiring small *kalata* (small water sources, often intermittent, which would support a few square meters of cultivation – usually fruit, vegetables, and alfalfa – and a small flock in the summer) in the mountains to the north. Every year about mid-June, when the harvest and threshing were over, all those who could would take off to the mountains with their flocks to avoid the heat and the malaria.

Barring disasters requiring major repairs, the qanat would continue to flow without maintenance for many years, though after a time the volume would probably diminish. The villagers tell of regular raids on their harvest and personal belongings by marauding Baluch from the south, who also ranged far to the north of Deh Salm and plundered much larger settlements. Forced to borrow in order to survive the year, the villagers gradually mortgaged and sold the title to much of their land, water, and palms, and stayed on as tenants on their own land. Because of the risks involved, it was not an attractive investment for large absentee landlords, who would fulfill at least some part of the function of the former khan. Instead, the land was sold to the people in the mountains, for whom the relatively lush property of Deh Salm had value despite its insecurity.

Finally, however, natural disaster also came. Sometime during the 1920s, after an unusually heavy rainfall, debris from runoff almost completely closed the qanat, leaving only a trickle. Without the injection of a major investment of labor – that the villagers were totally incapable of organizing – Deh Salm would have ceased to exist. The landowners in Neh were not interested. Finally, the khan of Birjand (200 kilometers to the north), the souzerain of the whole of southern Khorasan, was persuaded to reopen the qanat at the price of nine *meh* of the resulting flow of water (4½ days out of the 12½-day cycle).

After some years, the khan sold his share, and it eventually reached the hands of six brothers in Neh (about 1940). Perhaps the markedly increased security under the reign of Reza Shah (1925-41) had made investment in Deh Salm look more attractive again. These brothers financed more work in the qanat, and with the increased flow set about bringing more land under cultivation. They began to stake out land in what is now the southeastern tongue of cultivation, where the pottery and beads mentioned above have been found on it. Sense of the common identity of land and community, however, was sufficient for the villagers to rise up against what they considered to be the brothers' infringement of their territorial rights. A compromise was reached whereby land already staked out was left with the claimant, and the remainder of the cultivable land in that sector which could be irrigated was divided up among the community of shareholders in the qanat according to the amount of their share. Since that time some of the brothers have sold out, and others or their heirs have settled on one side of the village and become resident cultivators, performing some, if not all, of their own cultivation.

Until I became familiar with the situation in this village, I assumed that in the Persian situation a resident landowner was by definition a member of the village community. However, in Deh Salm the investing brothers and their descendants, whether or not they have become resident in the village, are not considered to be members of the community. The village community has retained its exclusive identity despite the decline of recent generations. Investment in the qanat, with or without residence, did not bring membership in the community.

## Nayband[3]

Deh Salm was essentially incapable of functioning without either (1) centralized authority or (2) a centralized investment system. The situation contains many features which are commonplace in qanat-based villages on the Iranian plateau – a region in which very few traditional agricultural villages were viable without a qanat. Nevertheless, some villages on the Iranian plateau depend on irrigation systems which are not so investment-intensive. The case of Nayband, approximately 150 kilometers northwest of Deh Salm, serves as an example of these. Nayband is also an oasis, but the two villages differ in almost every other respect.

Nayband has a larger population (around eighty families) and a smaller cultivated area. Instead of depending on a single source of water which represents a relatively major investment, it depends upon one major and several minor springs. These of course require no investment, and the channels which carry the water from them to the terraced fields require only a minimal amount. The village appears to have enjoyed a continuous existence since before the advent of Islam. Its economy is highly diversified, including besides agriculture and pastoralism, the trade of goods and services to passing caravans and migrant labor in mines. However, the village could not continue to exist without agriculture, and agriculture is impossible without irrigation.

In the Nayband irrigation system, the temporal and social subsystems are complex, but the economic subsystem is simple. Nayband is really a group of villages, arranged in the foothills and on the lower slopes around the base of a mountain of the same name approximately in the center of the deserts of eastern Iran. The name is applied to the whole community, the mountain, and the main village according to context. In what follows, the name Nayband is used to denote only the main village, where the great majority of the population resides. The remainder of the population is distributed in groups of one to ten families in settlements (*kalata*) around the mountain. The distance by foot around the mountain, taking in all the *kalata*, is approximately 60 kilometers. All the families in the *kalata* have title to property in Nayband, and all the dead are brought to Nayband and buried in one or another of the cemeteries there. Nayband and the *kalata* together, therefore, though spatially separate, form a single social universe or community. This community is neither politically nor economically stratified.

Most of the *kalata* depend upon small springs, and are from the point of view of irrigation, entirely separate from Nayband. One of them, however, Zardgah (or Ziyaratgah), shares the major spring of the region with Nayband, which is two kilometers away. Water from this spring, which accounts for some 90 percent of the cultivation of Nayband and all the cultivation of Zardgah, is distributed according to a ten-day cycle. The water of two consecutive days remains in the cultivated area of Zardgah. During the remaining eight days of the cycle it flows untapped along an open channel down a wadi to the cultivated area of Nayband. On the irregular and infrequent occasions that the wadi floods, the channel is completely erased in parts, but can be redrawn within a few hours.

The flow is divided into shares called *fenjan* (literally "cup") by the holed-bowl method previously described. There are 85 shares in one day's flow, making 850 altogether in the ten-day cycle. One share sells for 1,000 Rials, and the price is said to be rising. The day's flow is sufficient to irrigate twenty to twenty-five irrigation segments (here called *pella*), which are the minimal land parcels, and vary enormously in size because of the terrain.

The water flow is measured by the women, and for most of the cultivation it is measured into small ponds or tanks, each located above land owned by a distinct grouping of families whose shares occur together in the cycle. This empoundment allows it to be released on to the terraces with greater pressure.

All cultivable land in Nayband is named after the source of its irrigation water. Every possible square inch is prepared for cultivation by building terraces out from the hillside and up from the wadi bed. The creation of these terraces represents the most considerable investment involved in the system, but this investment is still within the capabilities of the individual operators. One sizable segment (*pella*) of the better land (perhaps 200 square meters in area) sold for 10,000 Rials and required two to three shares (*fenjan*) to irrigate it adequately, depending on its situation.

The major crops in Nayband are wheat, millet, and alfalfa. There is also a large number of date palms, but because no attention is paid to their cultivation, the fruit they produce is fit only for making syrup, a staple item in the diet, or for fodder. There appear to be two reasons for this lack of attention. First, the area was traditionally insecure, and the villagers never knew whether they would be able to harvest the trees. Secondly, however much attention they paid to cultivation, the village economy could probably never be viable from agriculture alone because of the quality and quantity of the soil and water. The area is climatically marginal for date cultivation. Even the wheat and millet is of very poor quality, and often fail altogether and are used for fodder. The alfalfa is, of course, cultivated as a fodder crop for flocks of goats which are grazed on and around the mountain.

In practice, the relationship between subsistence and the social subsystem in Nayband is complex. Although every member of the community owns shares, the nature of the resources are such that few if any own enough to subsist from agriculture alone. Traditionally, they have supplemented their income by catering to the caravan traffic (apparently including confidence trickery, smuggling, and outright robbery), but since the advent of the motor age, this traffic has disappeared. As a result, the adult men have had to look for other resources to supplement their subsistence, and the gap has mostly been filled by migrant laboring in the few small mining concerns operated in the deserts by the government and certain city-based investors. Work in the mines requires extended absence from the village and precludes agricultural work. The result has been that a varying proportion of the men conduct all the cultivation according to intricate individual agreements involving renting, sharing, and swapping, while the remainder of the shareholders tend the flocks or work in the mines.

In effect, therefore, the people of Nayband were primarily pastoralists, who irrigated largely in order to supplement by cultivation the insufficient natural pasture of the region. The single most important *raison d'être* for the settlement was probably its situation as an isolated, reliable watering point on the only caravan route which crossed the desert from Kerman in the southwest to Meshed in the northeast.

## CONCLUSION

In other studies, Deh Salm and Nayband would both be termed "small scale irrigation societies." However, in the terminology proposed in this essay, the irrigation system of the former (which incidentally has only half the population of the latter) is complex, while that of the latter is simple. This distinction allows us to predict – correctly – that the former contains elements of stratification and centralization, while the latter does not. Likewise, the criterion of size allows us to predict that because of the size of its annual product in relation to the amount of investment that its irrigation system is likely to need (albeit irregularly), Deh Salm cannot remain economically independent. I submit that this prediction is more obvious and less interesting than the first. However, since we have arrived at it, we may note that if the village of Deh Salm had been larger, it could have supported its own independent khan or large landowner. As it was, when the village lost its dependent khan, it was only a matter of time before it had to attract outside investment in order to survive. We may, therefore, deduce that the village must always have been a dependent colony – unless, as the ruins in the plain suggest is possible, it had once received sufficient capital investment and population to make it large enough to be economically and politically independent.

In the consideration of the simple system of Nayband, on the other hand, the criterion of size is actually irrelevent. Not only did Nayband not require any investment from outside, but there is virtually no way that an outsider could have invested in the system, unless of course modern changes were made in the traditional technology. Nayband appears always to have remained economically and politically independent and unstratified. The Khan of Tabas (historically, the nearest political center, 100 kilometers to the north) remitted Nayband's taxes in consideration of its function as a military outpost of his demesne, and welcomed the Naybandis individually when they had occasion to visit his court. One suspects that his magnanimity was hopeful rather than compensatory: it was not worth the effort to try to collect taxes from such a poor oasis, and remission was a cheap way to buy friends in a strategic situation.

These two small isolated villages, only 150 kilometers apart in the Iranian deserts, both dependent upon irrigation, could hardly be less alike. I have sought to show that the more significant sociological differences between them can be predicted from ecologically oriented studies of their technological systems. In such studies it is essential to proceed according to a typology and a set of analytical criteria which are derived from the same ecological approach, and not to adopt these criteria ready made from a different set of premises – as Steward did from Wittfogel.

## NOTES

[1]The field data referred to in this essay are from pilot ethnographic studies carried out in Nayband and Deh Salm June-August 1969 and June-August 1970 respectively. This research was made possible by grants and other assistance from the University Museum of the University of Pennsylvania, the Social Science Research Council, and the Office of Health, Education, and Welfare, to whom I am happy to record my gratitude.

[2]The Nile of course also provides examples of canal irrigation. Similarly, many of the systems cited contain examples of more than one of the types listed.

[3]The sketch of Nayband given here focuses on the operation of the irrigation system. Similar sketches written for different themes may be found in Spooner 1971 and 1972.

## REFERENCES


ADAMS, Robert McC.
1965 *Land Behind Baghdad: A History of Settlement on the Diyala Plains.* Chicago: Univ. of Chicago Press.

BEARDSLEY, Richard K., John W. Hall, and Robert E. Ward
1955 *Village Japan.* Chicago: Univ. of Chicago Press.

BOSERUP, Ester
1965 *The Conditions of Agricultural Growth.* Chicago: Aldine.

BRIGGS, Lloyd Cabot
1960 *Tribes of the Sahara.* Cambridge: Harvard Univ. Press.

BUJRA, A. S.
1970 *The Politics of Stratification: A Study of Political Change in a South Arabian Town.* London: Oxford Univ. Press.

CHI, Ch'ao-Ting
1963 *Key Economic Areas in Chinese History as Revealed in the Development of Public Works for Water Control.* New York: Paragon Book Reprint Division.

DROWER, M. S.
1954 Water Supply, Civilization and Agriculture. In *A History of Technology,* ed. by Charles Singer, E. J. Holmyard, and H. R. Hall, Vol. I.

ENGLISH, Paul Ward

1968 The Origin and Spread of Qanats in the Old World. In *Proceedings of the American Philosophical Society,* 112:170-81.

FERNEA, Robert A.

1970 *Shaykh and Effendi: Changing Patterns of Authority Among the El Shabana of Southern Iraq.* Cambridge: Harvard Univ. Press.

FLANNERY, Kent V.

1969 Origins and Ecological Effects of Early Domestication in Iran and the Near East. In Peter J. Ucko and G. W. Dimbleby, *The Domestication and Exploitation of Plants and Animals.* Chicago: Aldine.

FORDE, Daryll

1971 Ecology and Social Structure (Huxley Memorial Lecture, 1970). In *Proceedings of the Royal Anthropological Institute for 1970,* pp. 15-30.

FRUMKIN, Gregoire

1970 *Archaeology in Soviet Central Asia.* Leiden: Brill.

GEERTZ, Clifford

1963 *Agricultural Involution.* Berkeley and Los Angeles: Univ. of California Press.

GRAY, Robert F.

1963 *The Sonjo of Tanganyika.* London: Oxford Univ. Press.

HAMDAN, G.

1961 Evolution of Irrigation Agriculture in Egypt. In *History of Land Use in Arid Regions,* ed. Dudley Stamp. Paris: UNESCO, pp. 119-42.

HARTLEY, John A.

1961 The Political Organization of an Arab Tribe of the Hadhramaut. Unpublished Ph.D. dissertation. London School of Economics.

JACOBSEN, Thorkild and Robert McC. Adams

1958 Salt and Silt in Ancient Mesopotamian Agriculture. In *Science,* 128[3334]:1251-58.

KEES, Hermann

1961 *Ancient Egypt.* Chicago: Univ. of Chicago Press.

LAMBTON, Ann K. S.

1953 *Landlord and Peasant in Persia.* London: Oxford Univ. Press.

1969 *The Persian Land Reform.* Oxford: Clarendon.

LEACH, E. R.

1959 Hydraulic Society in Ceylon. In *Past and Present,* 15:2-25.

1961 *Pul Eliya.* Cambridge: The Univ. Press.

LESHNIK, L. S.

1972 Land Use and Ecological Factors in Prehistoric Northwest India. In *South Asian Archaeology.* Vol. I. Cambridge: The Univ. Press.

LE STRANGE, Guy

1905 *The Lands of the Eastern Caliphate.* Cambridge: The Univ. Press.

MILLON, René

1962 Variations in Social Responses to the Practice of Irrigation Agriculture. In *Civilizations in Desert Lands,* ed. Richard B. Woodbury, *Anthropological Papers,* Univ. of Utah, 62:56-88.

ORENSTEIN, Henry

1965 Notes on the Ecology of Irrigation Agriculture in Contemporary Peasant Societies. In *American Anthropologist,* 67:1528-32.

PARSONS, Talcott

1964 *The Social System.* New York: The Free Press of Glencoe [1951].

PRICE, Barbara J.

1971 Pre-Hispanic Irrigation Agriculture in Nuclear America. *Latin America Research Review*, 6:3-58.

RAIKES, R. L.

1965 The Ancient Gabarbands of Baluchistan. In *East and West*, 15:26-35.

SIROUX, M.

1959 *Les caravanserails de l'Iran.* Cairo: Institut Dominican d'Etudes Orientales.

SPOONER, Brian

1965 Kinship and Marriage in Eastern Persia. In *Sociologus*, 15:22-31.

1969 Politics, Kinship and Ecology in Southeast Persia. In *Ethnology*, 8:139-152.

1971 Continuity and Change in Rural Iran: The Eastern Deserts. In *Iran: Continuity and Variety*, ed. P. Chelkowski. New York: The Center for Near Eastern Studies and the Center for International Studies, New York Univ.

1973 *Population Growth: Anthropological Implications.* Cambridge: The MIT Press.

STEWARD, Julian H.

1936 The Economic and Social Bases of Primitive Bands. In *Essays in Honor of Alfred L. Kroeber.* Berkeley and Los Angeles: Univ. of California Press.

1949 Cultural Causality and Law: A Trial Formulation of the Development of Early Civilizations. In *American Anthropologist*, 51:1-27.

1953 Evolution and Process. In *Anthropology Today*, ed. Sol Tax. Chicago: The Univ. of Chicago Press.

1955 *Irrigation Civilizations: A Comparative Study.* Washington, D.C.: Pan American Union.

1968 Causal Factors and Processes in the Evolution of Pre-farming Societies. In *Man the Hunter*, ed. Richard B. Lee and Irven Devore. Chicago: Aldine, pp. 321-34.

TE-K'UN, Cheng

1959 *Archaeology in China, Vol. Three, Chou China.* Cambridge: W. Heffer.

WITTFOGEL, Karl A.

1938 Die Theorie der Orientalischen Gesellschaft. *Zeitschrift für Sozialforschung* 7.

1957 *Oriental Despotism, A Comparative Study of Total Power.* New Haven: Yale Univ. Press.

CHAPTER 5

# WATER CONTROL IN TERRACED RICE-FIELD AGRICULTURE IN SOUTHEASTERN ASIA

J. E. Spencer
*Department of Geography, University of California, Los Angeles*

Irrigation is traditionally associated with the seasonally dry and the perennially arid lands of the earth. The concerns of irrigators in such regions focus chiefly on three phases of the irrigation problem: water supply, conveyance to the crop site, and distribution within the irrigable area. Each phase involves its own distinctive technologies in which there is considerable variation in the separate regions of the earth. Diffusion has spread many of the technologies so that similar elements may be identified in widely separated parts of the earth. One notable aspect in irrigation of all dry regions is that much of the water is consumed in the processes of irrigation, so that the disposal of surplus water is not commonly a specific element in the operating technology. In long-irrigated dry land regions, the flushing of accumulated salts and the drainage of near-surface saline water sometimes require attention, but these remedial technologies are somewhat separated from the procedures of irrigation in the usual sense.

In the great flat lowlands of the earth, the processes of securing, conveying, and distributing water to irrigable lands may have had similar technologies, whether these be the ancient systems or those of modern time. Large-scale irrigation technologies and institutional systems probably must be examined in their own right apart from the small-scale systems that relate to local environmental situations and cultural conditions.

This paper presents a preliminary examination of the basic technologies that operate in the humid hill country of southern and eastern Asia, in which "irrigation" is an applied term relating to the particular system of water control employed on wet-field, terraced rice lands. Here many of the elements of technology that relate to traditional small-scale irrigation in the dry regions of the earth are absent, a different set of elements must be examined, and the term *water control* must replace the term *irrigation* in the definitional sense. These elements probably were not the operative features when the initial beginnings of wet-field cultivation were first worked out, but they became operative as the wet-field system evolved into a functioning body of technologies that enabled the spread of the wet-rice landscape onto lands of varied and irregular relief, and they remained the primary operative elements until the modern large-scale projects began to open up the great lowland regions of southern and eastern Asia. At present the basic technologies continue as the operative ones throughout the hill country regions.

Wet-field water control in the humid hill country of southern and eastern Asia occurs in regions of variable but distinct physical relief, in which there are natural and permanent streams of water flowing through small valleys and ravines. The development involves fitting crop fields into uneven and rough landscapes by means of building terraces. The water is already there, and fields go to water rather than water being brought to fields. The terraced fields are laid out empirically in groups on contoured patterns according to relief of the local land surface, but an uppermost terrace is usually placed close to a source of water. The development of terraced field groups is a time-consuming process that must be calculated in decades and centuries. A key element in the construction of a wet-field complex must always be the arrangement of discharge points through which water is returned to a stream channel. There is some downward percolation of water through the bottoms of new fields, but good long-used fields become relatively water-tight in the subsoil zones so that comparatively little water is lost or consumed. Physiological processes of crop plants consume some water, as does surface evaporation, but the critical element in wet-field irrigation is to control the movement of water across and over the field group in such a way that the discharge volume is always directly related to the intake volume.

If the local physical relief permits, an initial field group may be expanded upstream, downstream, or onto the flanks of valleys wherever terraces may be constructed. A small but round-headed valley containing a small stream that gathers the drainage flow from the uplands above the valley head is an ideal location in which to install a group of terraces, since the construction of individual field units can take place on either flank of the stream channel below the break in contour that marks the valley head. If terrace locations eventually extend to considerable distances from sources of water, it may be necessary to construct ditches to convey water to such marginal locations, since these late fields often are built in localities of rather strong relief and may be above fields constructed earlier. However, there is usually a limit to the number of fields in a given terrace grouping, and this limit will depend upon relief, local rainfall regime, and the physical problems involved in controlling water on the terrace group. Without removal of water from a terrace group, the accumulation of rainwater during prolonged or heavy rainstorms can create unmanageable flow volumes on the lower terrace members in any local group, a factor that restricts the manageable number of fields in any one group. It is true that some local landscapes appear to be continuous series of terraces, but the need for water control is such that any one "irrigation unit" handles the flow for a relatively small number of terraces. Therefore, an extensive total area of terraces in fact is controlled through a series of independent irrigation units tied closely to one or more stream channels in such a way that control of water may be managed for each unit independently.

The developmental sequence often begins with a very small group of fields and the installation of a water control unit adjusted to the immediate needs. In a good situation affording possible extension, other groups of fields and water control units may be created contiguously or nearby. Gradually the possible sites are filled in by other terrace groups, and a rather large acreage may accrue along a relatively small stream, since the "pass water through fields" system always returns water to a stream, unlike the situation in dry-land irrigation. Historically the wet-field landscape became a permanent cultural landscape of increasing area and water control complexity. There are a few areas in which abandonment has occurred over the last three thousand years, but the terraced wet-field landscapes have generally grown more extensive over the centuries. In the modern period there has been little significant addition to the wet-field landscape in the hill country, owing to the high costs of installing terraces on the remaining relatively unfavorable sites.

The operation of a terrace group through the use of a relatively small water-control unit does not guarantee permanence of individual fields, of course, and washouts do occur at the outer margins of individual terraces, chiefly through the failure of the earth or rock subsoil section at some point on a hanging wall below a field. The irregular delineation of field margins in long-cultivated localities clearly indicates the sites of past washouts, since the repairs often involve narrowing the field at the precise point of the washout in order to provide a more solid base for the hanging wall and the water-control bund. In areas of pervious subsoils, the exceedingly wiggly delineation of the field margins is an indication of past washouts. Sometimes a horizontal spread must be allowed between an upper and a lower terrace in order to maintain the basic support in the hanging wall. Such surfaces may often be left in thin soil covering a rocky lower surface and become grass-grown, but at other times a very tiny dry-field plot may be installed on the slope to be planted in some garden crop during the growing season. Such a procedure is often preferable to attempting to maintain a high hanging wall with the largest possible area in wet fields. When the upper margin of a higher field must be reduced in area, the possible enlargement of the lower field often would involve cutting into subsoil too far, so that the horizontal separation of fields and the maintenance of a wide sloping bund bulwark is the practical solution.

Not only must the wet-field irrigation system have its balancing water discharge, but there is an acute problem of adjusting the flow across terraces within any one group so that the movement of water downgradient will be well distributed. The maintenance of control during dry spells may dictate a given pattern of flow with few points of water discharge, whereas during a rainy period the volume of water that falls on the fields may require more numerous points of discharge in order to prevent the accumulation of too much water at any given point. The wet-field terrace system necessitates continued attention during rainy seasons and during cropping seasons. Although irrigation is certainly an objective of putting water onto the fields, the issue of water control becomes a more complex one of continuous management of the volume and pattern of water flow, features that usually are much more simply handled in the irrigation management in dry regions.

Several illustrations are provided to demonstrate the pattern of water control involved in wet-field irrigation. Because mapping one of the larger terrace groups is an arduous chore calling for large amounts of time and manpower, these shown are abnormally small terrace units. Figure 5.1 presents only one water control unit out of the four that make up the

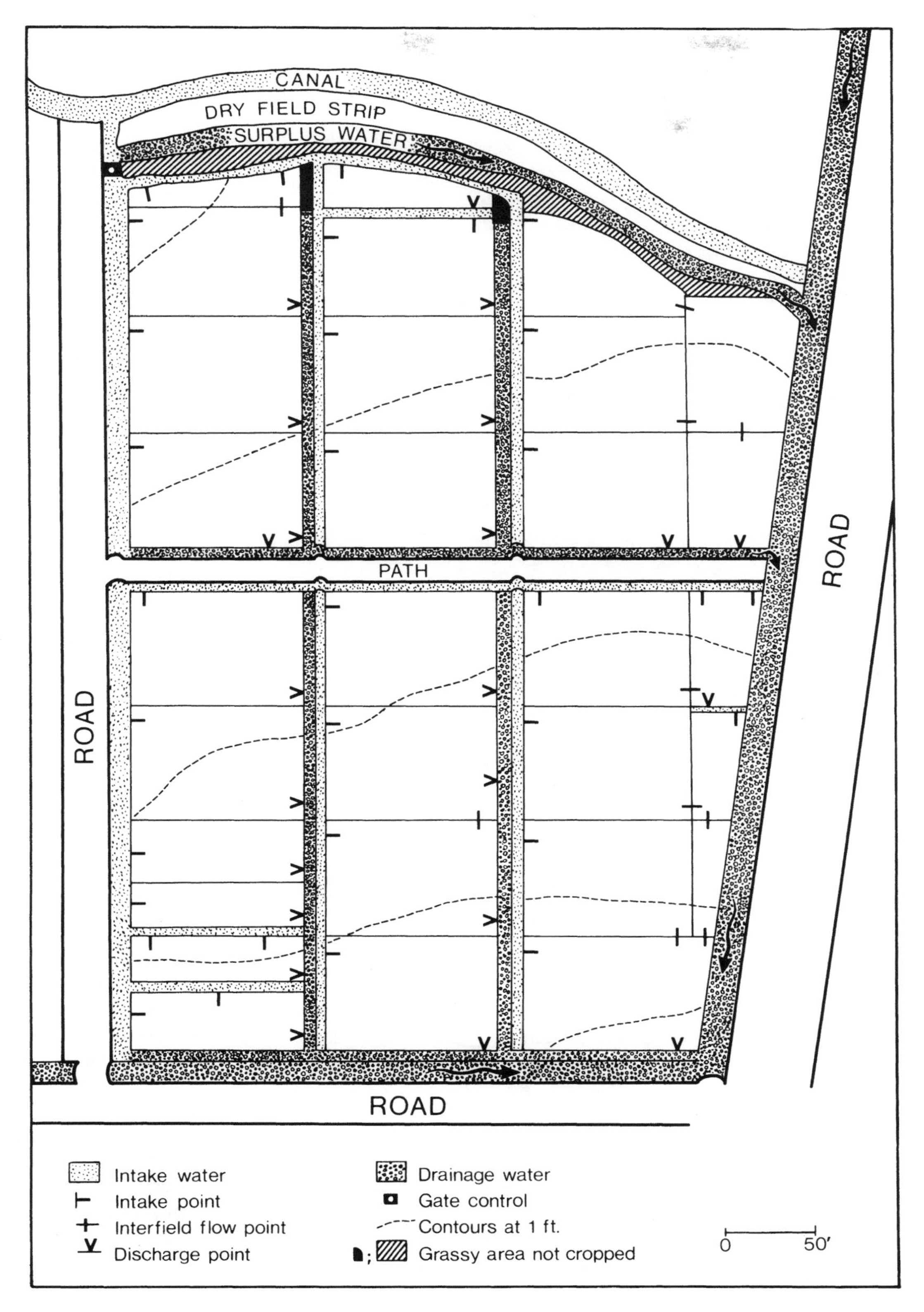


Fig. 5.1 Water control unit of the National Taiwan University experimental station. These fields are on a gently sloping plain, are bunded but only barely terraced, and are presented to illustrate in detail the degree to which control may be exerted over the gravity flow of water (May, 1970).

area of experimental fields worked by the National Taiwan University experiment station. This relatively flat and gently sloping site was chosen to provide a clear-cut example of the specific detail in the management of water flow on a quite flat surface. The water source for this field group is now a completely canalized stream that also provides water for other groups of fields downgradient. The separation of the inflow and the outflow into separate ditches in this specific example is not normal or common, but is a reflection of the care taken in an experiment station to maintain control of the total water supply.

Figure 5.2 presents the control system for a small group of terraced fields. It was constructed on a sloping natural bench along a valley flank having an antecedent primary stream on the left flank. This group of terraces is almost one hundred years old. On the lower right side-margin of this unit, the natural surface was cut and earth removed to level out the terraces above. There has been constructed an artificial consequent channel that takes the drainage off the upper field group and becomes a primary source of water to lower terrace groups. This small terrace group comprises all the wet-fields owned and operated by a five-family hamlet that also owned and cropped a larger area of dry fields nearby.

Figure 5.3 presents a somewhat different type of case, but one that is very common and found in the head of many a small valley. These fields are located southeast of Seoul. Above the terrace group, the stream flows in a narrow and rocky ravine. The first members of the terrace group were constructed at the uppermost point at which terrace building was then practical. The stream originally flowed in an irregular course roughly through the center of what now is the terrace group. When the second set of terrace members were installed in the upper center, and the original set of terraces were rebuilt, the stream's course was moved over against the steeper and rockier bank on which no terraces were practical. A heavy, thick, deep rock barrier was constructed at the primary point of diversion of the channel, and rocks in the original stream channel were transferred to the margin to provide a new rocky stream channel that would not suffer overdeepening by erosion. The original stream channel was then filled with earth scavenged from nearby localities and from the space which the new stream channel now occupies. This whole terrace group has accrued over about 150 years by additions on the left side (up slope), but a new highway constructed across the lower segment recently detached several fields that are now operated as a separate group and were not mapped. The youngest field members were installed during World War II (narrow fields on far left of this figure). The result is a new man-made landscape which has greater economic utility than the original one, but it is a landscape in ecological balance and one that very clearly simulates natural conditions.

Not every physical variant is illustrated in the three figures, but numerous others that have been mapped indicate that the basic elements are presented here. The implications of such irrigation systems for societal development would appear to be somewhat different from those of customary small-scale irrigation development in dry regions; they differ also from the implications of the large lowland systems. It is obvious that a large amount of labor is involved in the construction of terraces, but it is equally obvious that the labor has been expended by small groups over very long periods of time. No complex social organization is necessary to initiate, maintain, and expand hill country terraced fieldscapes, since their installations require no major structures and each small terrace group is, in effect, independent of neighboring small groups. Single families can begin such patterns of field development and water control, and families long resident in a locality can, and did, slowly expand the installation of added terraces. The wet-field system in the hill country margins of the Orient is susceptible to individual initiative among families or very small groups of families. This is not to say, of course, that every portion of the wet-field landscape of the oriental culture is the product of single-family piecemeal construction. Cohesive community groups can work at terrace installation in common and, within a relatively short span of years, can transform a relatively large wild habitat into a controlled cultural landscape. In many parts of the Orient such community action must have taken place on repeated occasions, but the significant aspect of such construction is that there result many separate and independent terrace groups symbiotically fitted into one or more drainage systems through the contemporaneous development of small units in independent operational sequences. Cooperation is obviously required between those operational groups, but that cooperation can be achieved at a local level without master planning or complex institutional sociopolitical controls.

In the traditional development of wet-field rice landscapes, cooperation among small groups resident on isolated farmsteads, in hamlets, or in small villages has been sufficient to facilitate initial development and continued expansion to the point of relatively full occupancy of given landscapes. It is common to find small terrace groups owned and operated by several families, and the terrace group presented as

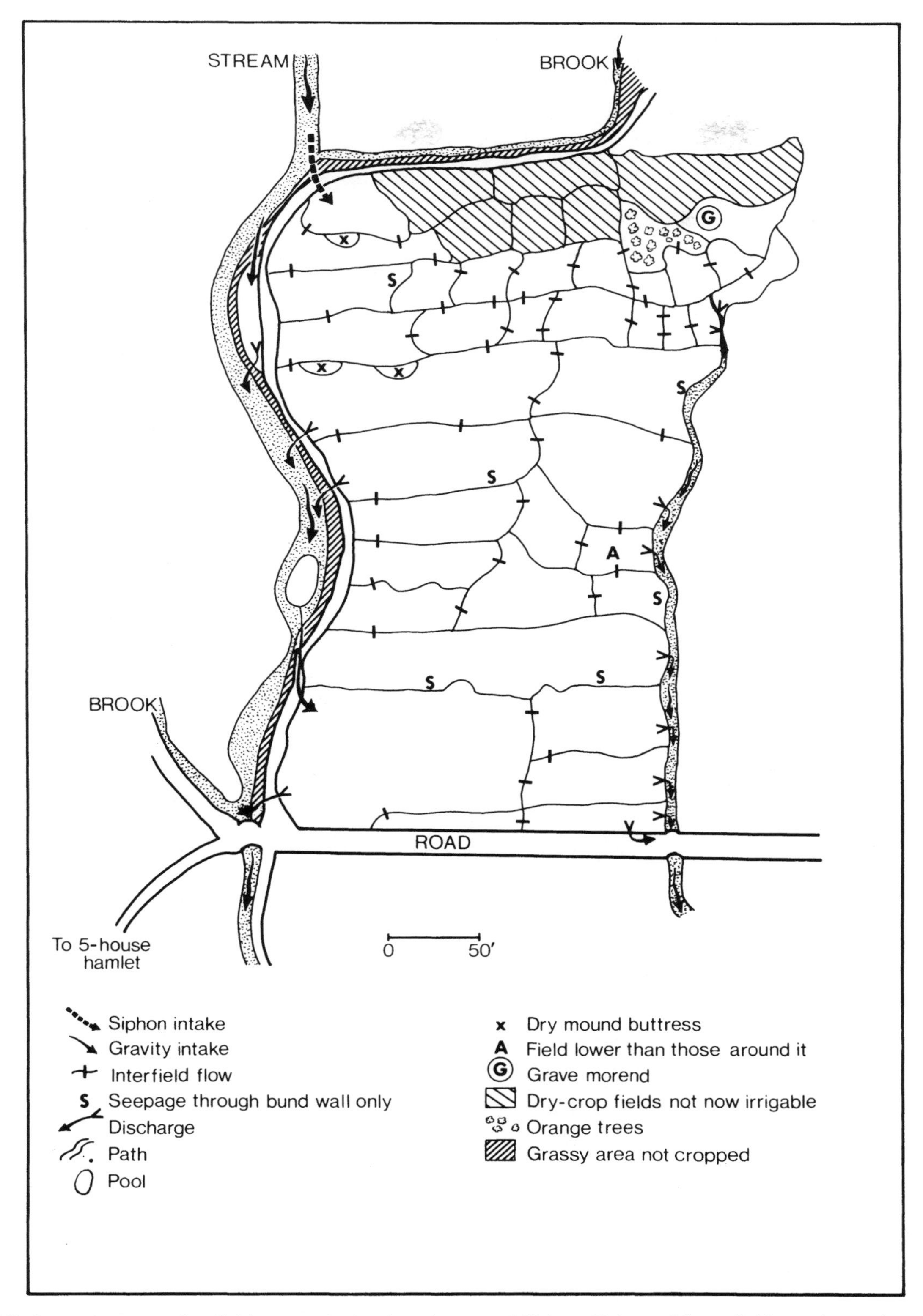


Fig. 5.2 Control system for fields on a sloping bench west of Taipei, Taiwan. These fields are located near the village of Ch'uch'ih, and were mapped in May 1970. The vertical drop from top to bottom on the left flank is ± 30 feet, that on the right flank is ± 34 feet, and the right lower corner is ± 50 feet below the left upper corner.

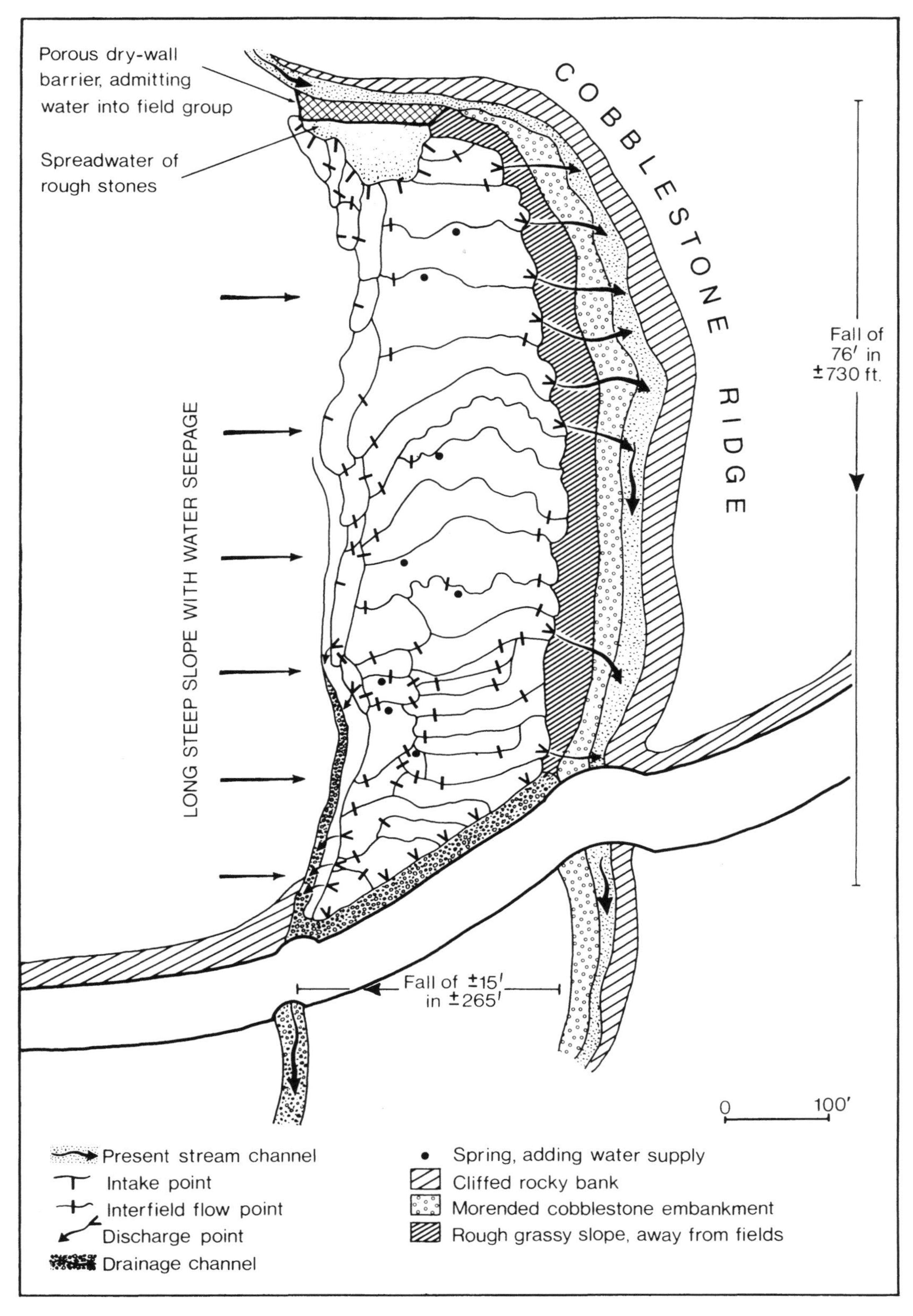


Fig. 5.3 Terraced fields southeast of Seoul, Korea. The original stream followed a channel not far from a median line drawn the length of the field group, so that the present landscape illustrates cultural control in the establishment and maintenance of a cultivated landscape (July, 1970).

Figure 5.2 is parcelled among seven families in a very mixed pattern of ownership of particular fields. The ownership of the fields presented in Figure 5.3 is also divided among seven families, but in this case the ownership units are segregated in contiguous units from top to bottom and along the upper left flank. Such local and essentially low-level cooperation seems both ancient and widely distributed in southern and eastern Asia among those culture groups that adopted sedentary living systems and permanent-field cultivation systems at some point in the past. It would appear that a critical factor in this whole development is that minimal level of understanding of the technologies-economics of permanent-field cropping systems. Given that minimal understanding and acceptance, the development of wet-field rice cropping, both as to terracing and as to water control, seems to have been a land-use system not calling for any particular sociopolitical organization that can be said to have been influenced by the water control system itself. There are many shifting cultivator societies in southeastern Asia that display varied complexities of social organization closely related to the patterns of their land-use system. Few such societies, within our historic record at least, have accepted permanent-field cultivation systems and discarded shifting cultivation systems; most modern attempts to initiate such technological transfers have met defeat for the occidental sponsors of the efforts. We know little about willingness or unwillingness to accept the permanent-field systems of land use and how such acceptance came about among culture groups in earlier eras.

It is quite clear that, once the permanent wet-field system is accepted, it does exert some influences on the societal living system, although at present these elements can be stated only in tentative terms. The need for staying with the water control problem day by day and year by year perhaps tends to increase the sedentariness of the families involved. Once the terrace groups are laid out, meaningful and effective cooperation is required between individuals holding fields within any water control unit. The lasting stability of a well-maintained assemblage of terrace groups makes for social and economic stability and for local autonomy in the political sense. One can speculate that the autonomous stability of wet-field rice-cropping terrace systems was a significant element in the evolution of the rural peasantry in a large sector of the oriental culture realm, but it cannot be asserted that such was the only factor in that evolution.

CHAPTER 6

# THE SYSTEM NOBODY KNOWS
## Village Irrigation in the Swiss Alps

Robert McC. Netting
*Department of Anthropology, University of Arizona*

Although the European Alps are not generally thought of as arid, in many areas dryness severely limits local agriculture and herding under natural conditions. This phenomenon is particularly apparent in Valais Canton of southern Switzerland (Fig. 6.1), where extensive mountain ranges to the south and north block moisture-bearing winds from the Mediterranean and the Atlantic. The upper valley of the Rhone from its glacier source to Lake Geneva is left in a rain shadow with only 50 to 80 centimeters (roughly 20 to 30 inches) of annual precipitation (Bär 1971:43). It is also relatively less cloudy than other regions of Switzerland, so that, for instance, the cantonal capital Sion has a yearly average of 400 more hours of sunlight than Zürich (Gutersohn 1961:13). Local valley winds blowing on fair days in summer may further increase transpiration.

The effects of continental climate are intensified on southward facing slopes (*Sonnenseite*) where exposure to the sun is highest. This condition is exemplified in Törbel, a German-speaking village in Vispertal.[1] Rainfall figures are unrecorded, but they may be slightly less than neighboring Grächen, with 56 centimeters, and Staldenried, whose 53 centimeters is the lowest average in Switzerland (Gutersohn 1961:53). Most of Törbel's territory is located on a southerly incline, receiving its major unobstructed insolation down the valley of the Mattervisp from the direction of Zermatt. A continuous, rather steep slope rises from the river at 770 meters, through the main village site at 1500 meters and the timber line at 2200 meters, to the peak Augstbordhorn at 2972 meters. The favorable exposure and the variety of accessible microenvironments based on altitude provide niches for all the major forms of subsistence – vineyards, grain fields, gardens, hay meadows, forests for fuel and building materials, and alpine pastures for summer grazing (Netting 1972). In the past, the community was largely self-sufficient in its staple dairy products, bread, and wine, its timber and stone for houses, and its wool for clothing.

The high location of the village, though it made transportation difficult, gave protection from the floods, malaria, and occasional marauding armies menacing the Rhone valley floor. Törbel is also well sited in regard to avalanche and rock slide danger. Such advantages, along with the presence of springs for drinking water and deep-soiled terraces for the planting of rye, characterize the mountain areas which were settled in pre-Christian times by Celtic populations (Staub 1944).

The crucial adaptive technique in this physical environment is irrigation. Without it, the rocky slopes above 1000 meters would carry only xerophytic (*Felsensteppe*) vegetation or the semi-arid larch forests from which Törbel takes its Gallic name (Zimmerman 1968:19-20). With intensive watering, the lower meadows can produce two lush hay crops plus a few weeks of grazing every year. Irrigation generally results in a four-to-five-fold increase in productivity over dry-farming.

Mountain irrigation is not rare in Europe, and indeed, Burns (1963) cites it as one of ten traits distinguishing a circum-Alpine culture area. Sections in the Alps have been irrigated for at least a millennium. Mariétan, in his richly illustrated essay (1948), presents historical documentation for irrigation in Valais from the eleventh and twelfth centuries. At present, canals of ten to twenty kilometers are common. A count in 1907 indicated that over 200 major channels, with a combined length of perhaps 2000 kilometers, were in operation (Stebler 1922:70-71).

In the simplest form of irrigation, water is merely diverted by a dam from a perennial stream and led through a stone-lined channel (*Wasserleitung*). Shallow feeder ditches (*Rus*), dug out of the soil, follow the contour of the meadow. Water is directed through

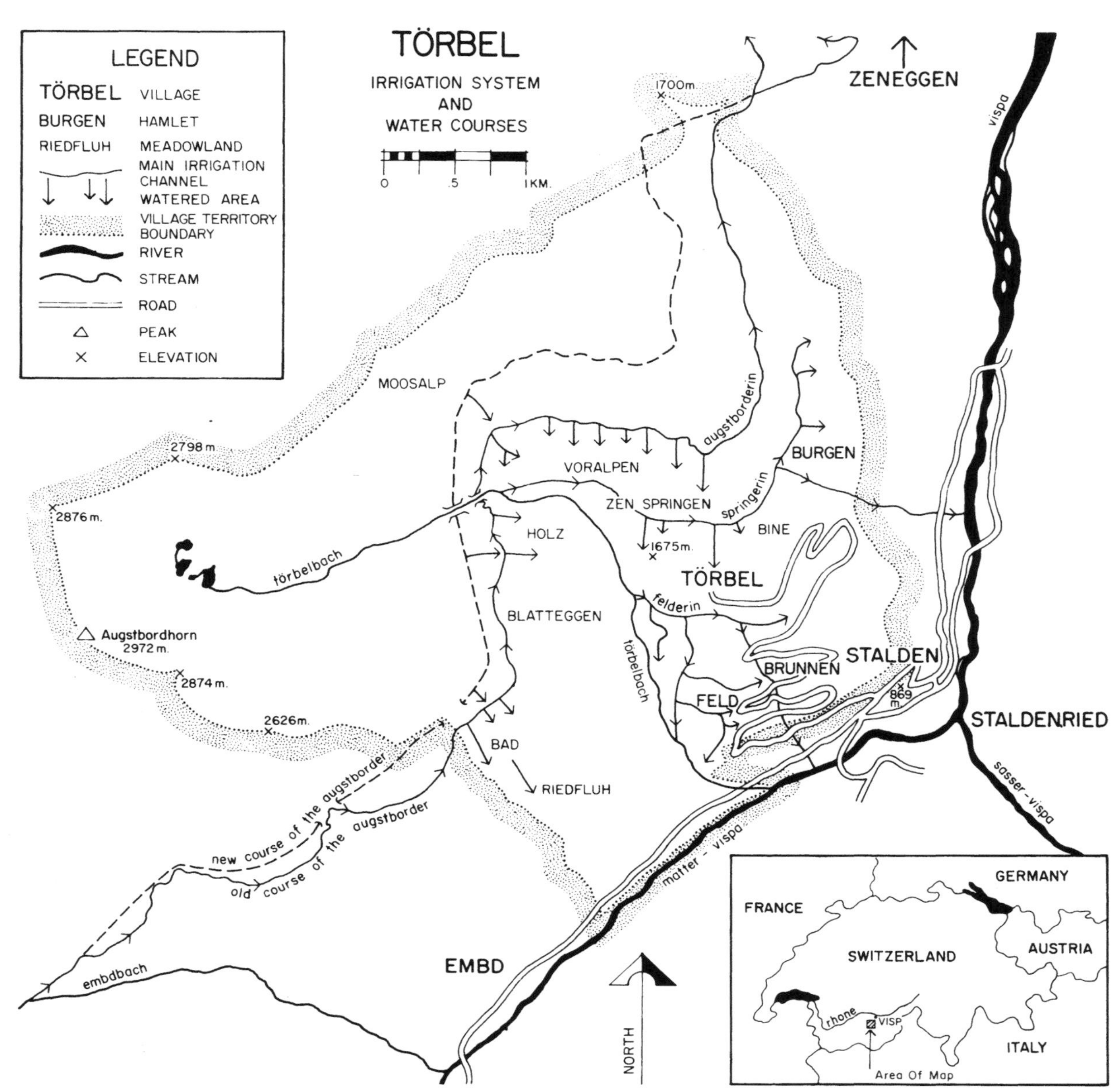

Fig. 6.1 Map: the Törbel irrigation systems.

the main branches of the system by wooden sluice gates (*Schieber*) or temporary earth and stone dams. Beginning at the downstream end, the feeder ditches are blocked at several points by stone plates (*Steinplatten*) thrown down with enough force to lodge in the earth sides of the trench. The partially interrupted flow spills out of the ditch and downhill. When the water reaches the base of the meadow, the stone plate is removed and placed a meter or two farther back to water the next strip of land. The same method is used to distribute the water of small spring-fed catch basins, which are drained several times a day as rapidly as they fill up.

Most Törbel farmers have meadows scattered at various altitudes and in widely separated parts of the village territory. A larger-than-average landowner might have water rights requiring irrigation of some plot almost every day of the week. To actual watering time of one to four hours, must be added a round trip of up to an hour and a half. Irrigation is most efficiently done by a team with one person (often a child) reporting the arrival of the water at the foot of the meadow. It is no wonder that one computation puts hours of irrigating at double the entire time spent mowing, raking, and transporting the hay produced (Stebler 1922:81). Watering chores also fall at the height of the summer agricultural season, when time is at a premium.

Törbel has three irrigation systems (Fig. 6.1). The two lower networks, the Felderin and the Springerin, are simpler, smaller, and very probably older than the uppermost Augstborderin. Both tap the Törbelbach, a brook which originates from snow-melt water and springs in a large mountain basin (Törbeltelli) on village territory. The Springerin takes two thirds of the brook's flow at approximately 1900 meters, while the remaining water is captured by the Felderin at 1500 meters to irrigate a wedge-shaped section of some 19 hectares just below the village. The highest tier of meadows (die Voralpen) and the outlying areas to the southwest and northeast are irrigated by a 10-kilometer canal called the Augstborderin or Niwe.

According to one account, water rights in the Embdbach were purchased from the valley town of St. Niklaus in 1270 by Törbel and the neighboring community of Zeneggen, which together constructed the channel and continue to share its water (Bichsel and Hämmerli 1967). Another source refers to a similarly ancient canal in much the same location which served five villages and was built by the Count of Visp. Though the origin of the system is unclear, documents indicate that it was independently controlled by an association of Törbel and Zeneggen residents in 1343 (Stebler 1922:71-72). The watercourse was rebuilt in the seventeenth century, in 1901, and most recently in 1947, when underground concrete conduits were installed at a cost of 2.2 million francs.

Each irrigation system was formerly managed and maintained by an association of users (*Geteilschaft*), whose annual labor contribution and infrequent cash assessments were in proportion to the water share of each member. The two lower systems are now nominally administered by the village council, while the long canal remains in the hands of an association, formally chartered, which elects officers, and is responsible for any necessary major financial expenditures. Water is allocated from the Augstborderin in a regular sequence, beginning at the downstream end and going to each successive section for a quarter-day (*Viertel*) of six hours. These quarter-days are used by one or more property owners with the order and time of individual shares listed precisely in hours and minutes. Traditionally the cycle (*Kehr*) was 21 days, with each community having approximately 42 quarter-days. With an increased water flow resulting from the last reconstruction, Törbel and Zenaggen use the water concurrently, dividing the volume in half. At various times, shares have been bought from one community by members of another. These purchased rights entitle the owner to receive water in terms of the geographical position of his meadow in the previously established sequence. Each cycle starts six hours later than the previous one so that watering times for an individual rotate regularly through the 24-hour period. Prior notice of the beginning of a cycle is given as one of the civil announcements in the brief public gathering after Sunday Mass. Supervision of the system is provided in yearly succession by one man (*Niventeiler*) from each group watering within the same quarter-day. In the past, the members from the two villages worked for a few days in the spring to repair breaks in the canal caused by fallen trees, avalanches, or gullying over the winter. A guard (*Wasserhüter*) was employed by the association to walk along the channel during each day of its operation to check for leaks and obstructions.

The two lower, older, and more spatially restricted systems are regulated without central authority or official mechanisms for adjudication. Upkeep is provided by a half dozen men working communally one day in spring, and the system is then left to run itself. Water rights are not publicly recorded, and no one knows even in outline the entire pattern of water distribution. Indeed informants universally deny the possibility of comprehending the total system. Each individual can supply information on when and where he is entitled to water, but no one can accurately list the shares for a whole day, much less an entire cycle. The usual phrase, accompanied with a vague, outflung gesture, is "Das Wasser ist bald her, bald da," meaning "the water is now here, now there."

The lack of definite knowledge and standardized administration contrasts with the elaborate codes of rules and detailed on-going records with which voluntary associations and the community as a whole regulate their activities. Written documents covering land ownership go back in some cases to the medieval period. Management of the common grazing ground on the alp and the division of cheese produced there require extensive listing of participants and quantification of their contributions. Vital statistics, cattle insurance, communal vineyard work, health plans, and the minutes of a multitude of voluntary organizations are all handled locally. Literacy is general and most adults can perform quite complex bookkeeping and bureaucratic chores. The Augstbord canal is managed with typical efficiency and organizational skill, though by a voluntary association which is unconnected with village political leadership. Why then are the two smaller irrigation networks so obviously lacking in centralized control and clearly stated rights in this valuable resource?

Unified knowledge and administration of irrigation is rendered difficult by the timing of shares and the complex rotation of individual turns in successive cycles. The following examples come from the Felderin, a system which I attempted to describe fully by collecting and collating bits of information from 33 members. For the Felderin, the standard cycle consists of 16 twenty-four hour days beginning the first Saturday in April and continuing, with the exception of Sundays, for every weekday through early September. The duration of water use on a particular plot is fixed with reference to the movement of sun and shadows on the surrounding landscape, e.g. *Dreifurren* (three terraces) is the time at which the sun first touches a particular field with three terraces above the village of Staldenried on the opposite slope, and *Schattigwasser* (shadow water) marks the moment when a shadow reaches a specified point on the Saaser Vispa River. The periods based on a variety of these *Wasserziele* (water markers) expand and contract according to the changing length of days and nights over the five-month season. The only fixed points are *Mittag* (midday), timed by the clock at 11:30 A.M. and *Mitternacht* (midnight), at 12:30 A.M. Table 6.1 lists the most common markers, the times at which they occur, and the average length of periods they define.

In describing his water rights, a farmer or his wife (women do a great deal of irrigating and may be better informed in these matters than men) notes first the day of the cycle, the distinctive name of the area where the meadow is located, and the markers defining each period through which the turn rotates. He may or may not know the named area (often not

**TABLE 6.1 Natural Markers Timing Irrigation Shares in Törbel***
**(Irrigation season for Felderi approximately April 11-September 4).**

| Marker | Approximate Times | | | Average Minutes in Period |
|---|---|---|---|---|
| | Earliest | Latest | Average | |
| Tagaufgang (Daybreak) | 2:50 a.m. | 4:15 a.m. | 3:30 a.m. | |
| | | | | 220 |
| Stadeldibschinu (Little granary shadow) | 6:30 a.m. | 7:45 a.m. | 7:10 a.m. | |
| | | | | 115 |
| Dreifurren (Three terraces) | 8:30 a.m. | 9:40 a.m. | 9:05 a.m. | |
| | | | | 145 |
| Mittag (Midday) | | | 11:30 a.m. | |
| | | | | 205 |
| Schattiwasser (Shadow water) | 2:10 p.m. | 3:45 p.m. | 2:55 p.m. | |
| | | | | 110 |
| Ottava (Place name in valley of Saaser Vispa) | 4:00 p.m. | 5:30 p.m. | 4:45 p.m. | |
| | | | | 50 } 140 |
| Ober Ottava (Upper Ottava) | 4:45 p.m. | 6:30 p.m. | 5:35 p.m. | |
| | | | | 90 } |
| Schattogspon (Shadow at Gspon hamlet) | 6:15 p.m. | 8:00 p.m. | 7:05 p.m. | |
| | | | | 325 |
| Mitternacht (Midnight) | | | 12:30 a.m. | |
| | | | | 180 |
| Tagaufgang (Daybreak) | 2:50 a.m. | 4:15 a.m. | 3:30 a.m. | |

*Times were taken from incomplete records kept by individual irrigators and rough records of times on the longest day in Stebler (1922).

contiguous) from which the water comes or the one to which it goes. A few meadows have fixed periods, recurring at the same time in each cycle. Most, however, rotate through a succession of periods, alternating at each time slot with one to six other meadows. The progression may be simple as in Table 6.2, where each meadow gets water in successive cycles at Period 1, 2, and 3. The initial order is determined by a representative from each meadow area drawing straws on the first Sunday in April. This order is then maintained throughout the summer, and some people make a note of it and mark the appropriate days on an almanac calendar when their water is due.

In any one meadow area, there may be several owners with varying proportions of the period, e.g. one-half, one-fourth, or two-thirds. Rotations may be more complicated as indicated by Table 6.3. Though Gapil Meadow follows a regular succession of periods, Bodman I and II divide each period and show an internal alternation, exchanging first position in successive cycles. Once the sequence is worked out in full, it accurately predicts irrigation times, but individual farmers find the schedule amazing because they reckon only the periods in which they are due to receive water. As patterns emerge in the analysis, gaps in time for which no user has been identified may suggest the presence of another meadow group, just as expected contrasts predict the presence of phonemes in a linguistic analysis. Table 6.4 shows a complex rotation in which one meadow (Ägerta) switches back and forth from day to night positions while Period 3 remains stationary, though this period is held by three meadows in successive years. Each day of the cycle shows different combinations and permutations.

Some flexibility is introduced into the system by allowing a member to use water during his allotted period for any meadow he owns in the area served by that irrigation network. Thus an inadequately

**TABLE 6.2 Felderin First Day.**

| Period | Meadow | No. of Owners |
|---|---|---|
| Tagaufgang – Stadeltibschinu (fixed) | Hogiblatt | 1 |
| 1) Stadel. – Mittag | Kapellenmatte | 3 |
| 2) Mittag – Ottava | Rosmatte | 3 |
| 3) Ottava – Tagauf. | Hubel, Sälli, Gufer | 3 |

**TABLE 6.3 Felderin Fourteenth Day.**

<table>
<tr><th>Period</th><th>Cycles: 1st</th><th>2nd</th><th>3rd</th><th>4th</th></tr>
<tr><td>Tagaufgang – Stadeltibschinu</td><td>Bodmen I</td><td rowspan="2">Hinter der Egge</td><td>Schluocht II</td><td rowspan="2">Gapil</td></tr>
<tr><td>Stadel. – Dreifurren</td><td>Bodmen II</td><td>Schluocht I</td></tr>
<tr><td>Dreifurren – Mittag</td><td rowspan="2">Gapil</td><td>Bodmen II</td><td rowspan="2">Hinter der Egge</td><td>Schl. I</td></tr>
<tr><td>Mittag – Schattigwasser</td><td>Bodmen I</td><td>Schl. I</td></tr>
<tr><td>Schattigwasser – Ottava</td><td>Schl. II</td><td rowspan="2">Gapil</td><td>Bodmen I</td><td rowspan="2">Hinter der Egge</td></tr>
<tr><td>Ottava – Schattogspo</td><td>Schl. I</td><td>Bodmen II</td></tr>
<tr><td>Schattogspo – Mitternacht</td><td rowspan="2">Hinger der Egge</td><td>Schl. II</td><td rowspan="2">Gapil</td><td>Bodmen II</td></tr>
<tr><td>Mitternacht – Tagaufgang</td><td>Schl. I</td><td>Bodmen I</td></tr>
</table>

NOTE: Bodmen I and II alternate in first position in successive cycles. Schluocht I and II alternate in first position in successive years (odd numbered year shown in table).

watered plot can be supplemented from water which is not needed in another area belonging to the same person. Neighbors may trade water, one taking two continuous shares in one cycle while the other does the same in the next. People who find watering at night unnecessary and inconvenient may lend such shares to someone else, often in return for help at plowing or harvest. Additional water may be purchased from the church, which has rights to all water on Sundays (*Sontagswasser*). One-hour periods are available on this day to landholders in order of their geographical arrangement. On main channels the sequence goes downstream on the right and then back upstream on the left. The series is again unrecorded and leads to lively arguments, which the church warden (*Kirchenvogt*) in charge of water sales makes no effort to resolve.

Given the agricultural necessity for water, its limited availability, and the sizable expenditure of labor involved in irrigation, one might expect either attempts to rationalize the system or frequent controversy. Neither of these occurs. The complex rotations through periods of unequal length build a self-correcting element into the system. Two owners (A and B) who attempt to take water at the same time may not have periods of equal length. If A takes the water, only to find that a third party (C) appears to claim it before the original period is over, or conversely that no one arrives to take the water from A at the end of his time, it is apparent that A is out of order. Those people whose periods are immediately affected can usually work out such problems among themselves, though the resolution may involve a recapitulation of all the cycles since the beginning of the season. Older farmers who are reputed to have a good head for these intricacies may be consulted by less experienced or confused neighbors. Dispute over the exact shadow marker to be observed is also settled privately. Everyone is aware that water is wasted when it must be led hither and yon, requiring time and the wetting of dry channels enroute, but it is deemed impossible to institute a regular progression of watering contiguous plots, as in the Augstbord system.

It is possible to steal water by leaving a sluice gate "accidentally" propped open or "forgetting" to remove completely a stone plate from a channel being used to convey water to another farmer out of sight in a down-slope meadow. Theft is difficult to prove, but if such apparent mistakes happen repeatedly, the injured party may give a warning and later resort to self-help, throwing the suspect's stone plates into the middle of the meadow. Harsh words may be exchanged, especially when water is in short supply, but informants do not remember more serious conflicts. The only local controversies known to have been brought to court are those in which ditches were

**TABLE 6.4 Felderin Third Day.**

| Period | Cycles: | 1st | 2nd | 3rd | 4th |
|---|---|---|---|---|---|
| Tagaufgang – Stadeltibschinu (fixed) | | Halbamatt | Halbamatt | Halbamatt | Halbamatt |
| Stadel. – 9:30 a.m. (fixed) | | Unter der Kapelle | U. d. K. | U. d. K. | U. d. K. |
| 9:30 – 10:30 (fixed) | | Feld bei der Kapelle | F.b.d.K. | F.b.d.K. | F.b.d.K. |
| 1) 10:30 – Mittag | | Ägerda | Feld II | Lengseich | Feld I |
| 2) Mittag – Schattigwasser | | Feld I | Lengseich | Ägerda | Feld I |
| 3) Schattigwasser – Schattogspo | | Boden | Boden | Boden | Boden |
| 4) Schattogspo – Mitternacht | | Feld II | Ägerda | Feld I | Lengseich |
| 5) Mitternacht – Tagaufgang | | Lengseich | Feld I | Feld II | Ägerda |

NOTE: Position 3, which remains stationary, is occupied in successive years by Boden, Feld I, and Feld II.

poorly maintained or water was allowed to flow unsupervised (*herrenlos*) through the system until it broke into a meadow and soaked the earth, causing a mud slide. The individual or association responsible can be sued for damages. Litigation over water rights or resort to higher authority is almost wholly confined to intercommunity relations. Törbel has had centuries of controversy with neighboring Embd over the amount of water which each draws from the Embdbach and over the maintenance of the Augsborderin, which crosses Embd territory (Imboden 1956:34, Stebler 1922:72).

There are several possible explanations for the development and preservation of an acephalous system of ordered anarchy in irrigation. The natural period markers obviously predate the clock, and the division of water may reflect a system which grew by gradual accretion over time. Little organized community activity was required to build the main channels, and as individuals extended the ditches into new meadow areas, they worked out limited and idiosyncratic agreements for water sharing. Though water rights accompany land, they are seldom specified in the elaborate deeds of land transfer that appear from the seventeenth century onward. It is reasonable to assume that inheritance and sale have subdivided properties and at times led to exchange and limited reapportionment of water rights. Some meadows with evidence of ditches are no longer entitled to water. A rationalized system of water sharing is resisted by those who derive advantage from the current arrangement. Convenient watering periods during the day are valued, and owners are reluctant to accept other times. Though everyone recognizes that some unfairness of distribution is perpetuated by the existing system, larger owners claim (1) that their water is as much a possession as the land and is subject to similar inequalities in tenure, (2) that any reorganization would be dreadfully complicated, and (3) such a project would inevitably arouse suspicion and animosity in all concerned. The most qualified individuals say they would refuse to plan a reorganization, and that successful change could only be brought in with the installation of a pipe and sprinkler system that would radically alter the amount and nature of work involved in watering.

Preliminary calculations of the holdings of nine farmers in the Felderin system support the contention that there are gross inequalities in amount of water available per unit of land. Table 6.5 shows that some plots have a minute of water for 3.7 square meters of area, while the same amount of time must serve for an average of almost 50 square meters in other meadows. The legacy of centuries of private deals and tinkering with details of the system has resulted in a variety of vested interests. Frequent movements of water from one area to distant parts of the network, reliance on natural periods of changing duration, and a multitude of incomparable rotation orders within the cycles all join to obscure the nature and inequities of the system. The lack of standardized public records and central control prevent any individual from enlarging his own circumscribed practical knowledge. It is easier to make up a lack of water by shifting supplies among plots, buying from the Church, and bargaining with neighbors and kin than to attack the system and its entrenched supporters.

From the Törbel case, it would appear that small-scale irrigation systems[2] which grow essentially without plan and are maintained with a minimum of cooperative effort can function on the basis of an intricate series of water-sharing agreements, each meshing with the others but known to individuals only insofar as their own use rights are exercised. When geographical factors limit the amount of arable land and/or water, and when cultivators control their own water source, be it a well, an impoundment, or a stream, there is no possibility for expanding the system. The association of users has no need to be coercive if its members' rights are conditional on obligatory contributions of labor and capital. Central control and direction of work is unnecessary if upkeep is minimal and routine. Emergency demands for work are self-evident, and the sanction for participation is simply loss of the water on which the individual's own agricultural success plainly depends. Experienced members of the association all have whatever engineering knowledge is required to keep the system functioning, and they are able to act as ad hoc or elected leaders when circumstances demand cooperation.

A system in which no one possesses comprehensive and comprehensible knowledge of its total operation has a kind of organic stability. Mistakes in order or efforts to take a larger share than one is entitled to attract notice and encounter immediate resistance from those whose turns are most closely integrated with one's own. Regulation and self-correction are paradoxically more effective when the parts of the system are *not* interchangeable and their interdigitation is an impenetrable maze. As Leach (1961:165) pointed out with respect to Pul Eliya's traditional water distribution, "the complexity of the arrangement is itself relevant. . . since such a system is virtually unalterable." A higher authority is not only

Table 6.5 Relation of Irrigated Meadow Area to Duration of Water Supply.

| Farmer | Day | Location | Area (sq. meters) | Average Time (minutes) | Square meters irrigated per minute of water |
|---|---|---|---|---|---|
| A | 1st | Kapellenmatte | 1190 | 77.5 | 12.8 |
| | 5th | Hausmatte | 1200 | 25 | 48.0 |
| B | 5th | Feld | 1240 | 25 | 49.6 |
| | 5th | Reft | 2000 | 336 | 5.96 |
| | 11th | Börter | 670 | 45 | 14.9 |
| C | 3rd | Ägerda | 1500 | 48 | 31.25 |
| | 13th | Stahlschier | 850 | 102 | 8.3 |
| D | 1st | Kapellenmatte | 530 | 102 | 5.2 |
| | 4th | Kapellenmatte | 360 | 20 | 18.0 |
| | 5th | Blattmatte | 110 | 30 | 3.7 |
| | 8th | Hublen (Feld) | 900 | 240 | 7.5 |
| | 6th | Hogiblatt | 3020 | 112 | 18.0 |
| E | 1st | Hogiblatt | 2750 | 220 | 12.5 |
| | 16th | Schluocht | 1450 | 360 | 8.1 |
| | 7th | Boden | 1041 | 30 (60 every other kehr) | 34.7 |
| F | 12th | Bodmen | 1105 | 85 | 13.0 |
| | 14th | Bodmen | 1050 | 56 | 18.75 |
| G | 4th | Bodmen | 1660 | 252 | 6.6 |
| | 3rd | Boden | 1580 | 128 | 12.3 |
| | 5th | Feldhalmerin | 320 | 90 | 3.6 |
| H | 2nd | Krumenstückini | 3000 | 92 | 32.6 |
| I | 14th | Hinter der Egge | 1330 | 100 | 13.3 |
| | 9th | Hofmatten | 1150 | 94 | 12.2 |

unneeded but actually irrelevant to the settlement of disputes about a distribution based on shadowy points of reference, eccentric rotations, and rigidly uncodified rights. In such a system, an uninformed, voluntarily cooperative association of users would seem the most appropriate social institution.

The tendency toward rationalization of water distribution, clarification of rules, formalization of rights, and centralization of powers may appear under the following circumstances: 1) when the source of water is not locally controlled, and rights in it must be purchased, rented, or acquired by force and thereafter defended from other communities competing for the same resource, and/or 2) when the building and maintenance of the canal system require the joint efforts of several otherwise independent communities. These factors are evident in the uppermost Törbel irrigation system, which relies on a distant water source and is tapped with the cooperation of a neighboring village. Conflict within the community of irrigators appears small in proportion to cooperation. Disputes, litigation, and hostility characterize relations between competing settlements, but even here, there is no evidence of the identification of control of irrigation and political leadership. The hardy Alpine variety of hydraulic society, even with the introduction of order and rules, must still be classified as an occidental democracy.

## NOTES

[1] Field research in Törbel from July 1970 to August 1971 was supported by a fellowship from the John Simon Guggenheim Memorial Foundation, a grant from the University Museum of the University of Pennsylvania, and National Science Foundation Grant No. GS-3318.

[2]"Small scale" is less a matter of relative size of system or amount of water distributed than of the investment of labor and resources necessary to construct and maintain a particular irrigation network, as Spooner (this volume) points out. "Scale" may also refer to the limited number of participants involved in water sharing and the presence of numerous cross-cutting ties of kinship, marriage, residence, common ritual activities, and cooperative economic endeavors which promote informal methods of dispute settlement.

## REFERENCES

BÄR, Oskar
1971 *Geographie der Schweiz.* Zürich: Lehrmittelverlag des Kantons Zürich.

BICHSEL, Ulrich and Reinhard Hämmerli
1967 Die Wasserfuhren. In *Torbel Studienwoche.* Staatliche Lehrerseminar: Bern-Hofwil. (Mimeo)

BURNS, R. K.
1963 The Circum-alpine Culture Area. *Anthropological Quarterly* 36:130-55.

GUTERSOHN, Heinrich
1961 *Geographie der Schweiz.* Alpen, Band II, 1. Teil. Bern: Kümmerly-Frey.

IMBODEN, Adrian
1956 *Die Produktions- und Lebensverhaltnisse der Walliser Hochgebirgsgemeinde Embd und Möglichkeiten zur Verbesserung der gegenwartigen Lage.* Schweizerische Arbeitsgemeinschaft der Bergbauern, Nr. 40. Brugg: SAB.

LEACH, E. R.
1961 *Pul Eliya: A Village in Ceylon.* Cambridge: Cambridge Univ. Press.

MARIÉTAN, Ignace
1948 *Heilige Wasser.* Schweizer Heimatbücher, Nr. 21/22. Bern: Paul Haupt.

NETTING, Robert McC.
1972 Of Men and Meadows: Strategies of Alpine Land Use. *Anthropological Quarterly* 45:132-44.

STAUB, Walter
1944 Der älteste Siedlungen in Gebiet der Vispertäler. *Neue Züricher Zeitung*, Nr. 2052, November 20, 1944.

STEBLER, F. G.
1922 *Die Vispertaler Sonnenberge.* Jahrbuch des Schweizer Alpenclub. Sechsundfunfzigster Jahrgang. Bern: Verlag des Schweizer Alpenclub.

ZIMMERMAN, Josef
1968 *Die Orts- und Flurnamen des Vispertales im Wallis.* Zurich: Juris Druck Verlag.

CHAPTER 7

# ORGANIZATIONAL PREADAPTATION TO IRRIGATION
## The Evolution of Early Water-Management Systems in Coastal Peru

M. Edward Moseley
*Peabody Museum, Department of Anthropology, Harvard University*

Coastal Peru provides an excellent opportunity for the study of prehistoric irrigation systems. Vast tracts of desert were once brought under cultivation by inhabitants of the coastal valleys. With the Spanish Conquest there was a decrease in the amount of land under irrigation, and many areas farmed in pre-Hispanic times have yet to be reclaimed. Therefore, extensive systems of ancient fields and their associated canals survive in desert areas beyond the margins of modern agricultural lands. These pre-Columbian survivals have been dated no earlier than c. 400 B.C. (the beginning of the Early Intermediate Period) are the products of complex societies with well-developed irrigation technology.

The surviving water-management systems postdate by more than two millennia evidence of early plant cultivation on the coast. Since the use of domesticated plants appears well before the earliest surviving irrigation works, and these works are of a sophisticated nature, it is logical to assume that water management technology had undergone considerable development elsewhere prior to the first appearance of canals and fields along the coast.

For the period of c. 1400 to 400 B.C. (the Early Horizon) there is as yet no direct evidence of coastal water-management systems. However, in most valleys that have received careful study, there is a heavy inland occupation generally associated with extensive and elaborate monumental architecture. These ceremonial centers and public works are viewed as labor products of substantial numbers of people. Almost all Andean archaeologists consider this pattern of inland residence and building to have been based on irrigation agriculture. Prior to c. 1400 B.C., the history of water-management technology is more difficult to reconstruct.

Concerned with the early development of water-management technology, this study first provides a review of data from the central coast that suggest what the course of development may have been, and secondly examines certain background conditions that may have fostered rapid adaptation and development of canal irrigation.

### THE COURSE OF DEVELOPMENT

Beginning about 2500 B.C. (the opening of the Cotton Preceramic Period), hunting and gathering were abandoned as a way of life on the Peruvian coast, and man turned to the full-time exploitation of marine resources. The coastal waters of Peru support one of the world's richest aquatic biomasses (Murphy 1923; Sverdrup *et. al.* 1946). These waters provided a set of resources extremely abundant, available year-round, and easily exploited. All present data indicate that the littoral-based economy supported marked population growth (Moseley 1972), and the founding of very large, permanently occupied coastal settlements (Lanning 1967). In terms of size, many of the Peruvian sites are thought to have housed populations equal to, if not considerably larger than, maritime settlements on the Pacific coast of North America.

Prior to the abandonment of hunting and gathering, man utilized and presumably cultivated gourds (Lanning 1963) and perhaps squash or beans. With the change to fishing and sedentary life, the repertoire of cultivated plants expanded slightly to include, among other things, legumes and domesticated cotton and peppers, as well as certain fruits (Patterson and Moseley 1968). However, none of the early plants was suitable as an agricultural staple (Pickersgill 1969). Where the food remains from middens have been given quantitative study, cultivated plants were relatively rare and always of secondary economic importance to comestible marine products (Engel 1963; Moseley 1968; Wendt 1964).

Thus, both the nature and frequency of early cultivated plants point to their having a minor role in subsistence. This interpretation is corroborated by settlement patterns of the period. Inland sites that might have served as farmsteads are rare or absent, and a majority of the recorded settlements, including two of the very largest, are situated away from water in arid sections of the coast where plant tending would be impossible (Engel 1958; Moseley and Willey 1972).

Because cultivated produce was of minor importance, it is generally believed that canal irrigation of desert lands was not practiced at this time (Lanning 1967). The assumption is that agricultural output – either for lack of capability or lack of need – did not justify an investment of labor in irrigation. Rather, plant tending is postulated to have occurred in river plains by means of floodwater farming. Although direct evidence of floodwater farming is lacking, this postulate fits closely with the nature of the plants in use, their midden frequency, and the settlement patterns.

Some time before 1800 B.C., but presumably after 2000 B.C., maize appeared in a limited number of maritime settlements on the northern part of the central Peruvian coast (Moseley and Willey 1972). Quantitative analysis of the associated midden deposits has yet to be carried out, but there is currently little to suggest that maize affected the secondary role of agriculture or immediately modified the existing subsistence pattern to a significant degree. In fact, the absence of this cultigen in either contemporary or somewhat later sites to the south points to a relatively slow diffusion for the use of this plant (Moseley 1968; Engel 1963, 1966; Gayton 1967). Thus, present information does not suggest that the introduction of maize carried immediate economic consequences.

Around 1750 B.C. (the end of the Cotton Preceramic Period and the beginning of the Initial Period), there was a widespread and major change in settlement patterns. Many, if not a majority, of the early fishing sites were abandoned. There are no indications that this abandonment was precipitated by a marine cataclysm or other problems affecting the maritime subsistence economy. There are also no indications that the surviving coastal settlements absorbed the displaced populations.

To understand what may have transpired, it is necessary to examine the archaeological record of the Ancón-Chillón-Rímac region (Fig. 7.1). This region includes the Rímac Valley where Lima is located, the adjacent Chillón Valley, and the large isolated desert bay of Ancón. There were four maritime settlements in the region, including the biggest early coastal site yet discovered. This site, El Paraiso, covers more than 50 hectares. It is situated near the mouth of the Río Chillón, where the inhabitants could engage in both fishing and flood water farming (Engel 1966). The other maritime sites are in desert areas.

Coincident with the period of widespread coastal abandonment, El Paraíso and one other local site ceased to be occupied. Because these two sites, taken in combination, were considerably larger than the two surviving fishing settlements, it is assumed that a majority of the local population has been displaced from the coast.

Around 1700 B.C., an immense platform mound, the Huaca La Florida, was built in the Rímac Valley some 14 kilometers up-river from the valley mouth. For its time period, this structure is the largest yet found in the Andean area. If it was built by local labor, as seems likely, a substantial inland population is implied. At a later date, another large platform was constructed away from the coast on the south side of the Chillón Valley. Again, if built by local labor a sizable valley interior population would be required.

These lines of evidence indicate that the abandonment of coastal settlements was part of a process involving the shift to inland valley residence by a

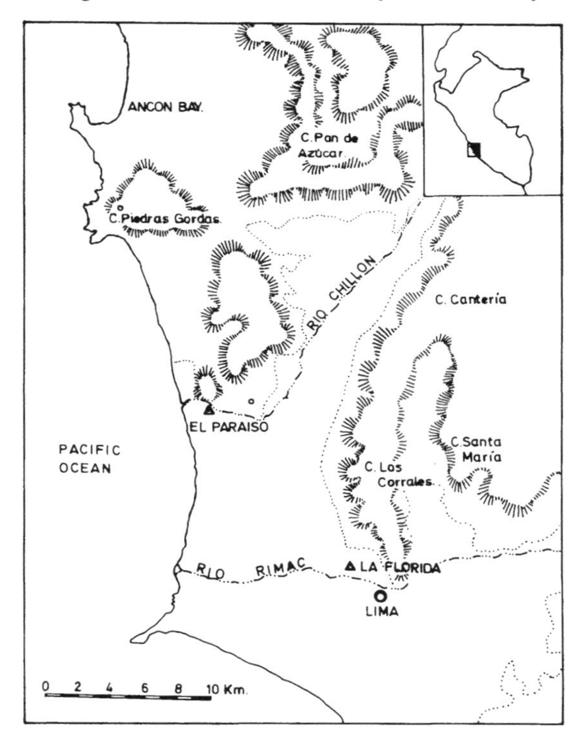


Fig. 7.1 The Ancón, Chillón, Rímac region of coastal Peru.

major segment of the local population. If such is the case, intensive use of agriculture is implied because, in the Peruvian desert, inland settlement by more than a small number of people must be predicated upon plant cultivation. There are no other economic options open to a preindustrial population of any size. The argument here is that the shift from coastal to valley interior residence reflects a shift in subsistence patterns, with agriculture displacing fishing as the primary means of support for a majority of the local population.

Countenance is given this argument by excavations in the area of Ancón Bay. Here the archaeological sequence shows that cultivated plant foods played a secondary and minor economic role up to the period of coastal abandonment. Immediately thereafter, both the type and frequency of domesticates in use at Ancón increased markedly (Moseley 1968; Patterson and Moseley 1968). In other words, there is clear evidence that the shift to inland residence was accompanied by a rise in the level of agricultural output.

It is possible to argue further that the valley interior occupation and the increase in farming productivity reflect the abandonment of floodwater farming and the introduction of canal irrigation, which opened desert lands to cultivation. The entrenched coastal rivers of Peru offer only a limited amount of land suitable for flood water farming. The restricted availability of land and the fact that flood water farming could produce but one annual harvest set tight limits on the level of agricultural output. On the other hand, canal irrigation could open vast tracts of desert to agricultural exploitation, and with sufficient water it would be possible to farm year-round. Thus, a change in water management technology that brought more land into use and extended the farming season could be expected to significantly raise the level of agricultural output.

The abandonment of El Paraíso at the mouth of the Chillón Valley supports this contention. The site is situated adjacent to an exceptionally large area of river plain suitable for floodwater farming (Engel 1966). Depopulation of the site and the apparent lack of reoccupation of the area imply that this tract of flood plain ceased to be an important economic resource. This can be construed as a result of the opening of more productive desert lands to cultivation.

In summary, present data on settlement and subsistence patterns from the Ancón-Chillón-Rímac region indicate that canal irrigation began c. 1750 B.C. (the beginning of the Initial Period).

The change in residence patterns which accompanied the use of this water management system is best understood in terms which postulate that canal irrigation was initially used in the upper and middle sections of coastal valleys, and not near the valley mouths. This hypothesis rests on the fact that there are sharper gradients in the upper and middle valley areas than in a valley mouth. This fact is significant for three reasons. First, in areas of steep gradients, lead-off canals needed to channel water out of the entrenched rivers can be relatively short. Near the coast, where there is little slope, the lead-off systems must be long – often in excess of several kilometers – and a large labor investment is required (Moseley 1972). Second, once channelled onto arable land, it is easier to direct water on inclined areas than on relatively horizontal surfaces. Third, in those areas close to coastal valley mouths, irrigation may result in water-logged and salinated soils, if there is not a substantial labor investment in the creation of proper drainage systems (Ian S. Farrington, personal communication 1972).

Therefore, if canal irrigation was initially practiced in upper valley sections, and if this system of water management fostered a rise in the level of agricultural output, the adoption of an inland settlement pattern is probable.

## THE BACKGROUND

If the evolution of early water management systems has been correctly traced, then an accelerated development of canal irrigation is implied. Fishing was the economic mainstay of coastal societies at 1800 B.C. By 1600 B.C., irrigation had opened the desert to intensive exploitation, and farming was the primary means of support for a majority of people. Thus, a major change in subsistence patterns transpired in less than two centuries, and irrigation played an important role in bringing about this change.

It is difficult to calculate exactly how fast canals were initially pushed into the desert. The apparent radical alteration in settlement patterns that occurred around 1750 B.C. may denote the opening of considerable land in a short time, and thus would imply substantial canal construction at an early period. In other words, the archaeological record does not give the impression of gradual development in the initial stages of irrigation technology. Rather, it points to the rapid building and use of canals.

It can be argued that irrigation of the desert underwent an early period of accelerated development because the coastal societies of Peru were in some ways preadapted to the use of this water management system. At the technological level, the desert peoples had over a millennium of experience with plant cultivation prior to engaging in irrigation agriculture. Therefore, the general tenets of farming had been worked out prior to the advent of intensive agriculture. Thus, the use and development of irrigation technology was not restrained by the necessity to formulate the principles of plant tending; these principles were pre-existing.

At the organizational level, the building of canals postdates considerable experience in corporate labor construction. Under the aegis of the maritime economy, people were living in large, nucleated settlements, the biggest of which are variously estimated to have housed 500 to 1,500 or more individuals (Lanning 1967; Engel 1966). Evidence for corporate labor undertakings appears at almost all of the more extensive maritime settlements, and there seems to be a general but imperfect correlation between site size and the scope of organized labor activity.

The nature of early corporate labor endeavors was variable. Both residential and non-residential construction were undertaken. At Huaca Prieta, a site of moderate size, group labor was used to contain and stabilize the summit of the settlement with large masonry walls (Bird 1948). Culebras I is a very substantial site situated on a hillside at the mouth of the Culebras Valley. Here, corporate construction entailed building masonry-fronted terraces over an area of many hectares to provide level residential surfaces (Lanning 1967). El Paraíso, near the Río Chillón, represents the most extensive of the early ventures in mass construction. The site consists of a large, thick-walled masonry complex of residential quarters that spreads over more than fifty hectares (Patterson and Lanning 1964; Engel 1966).

Corporate construction that was not residential in nature was undertaken at two large settlements. Río Seco, north of the Chancay Valley, is an isolated inter-valley site with two artificial platform mounds. In part, these were built with small boulders carried to the site from a source area more than one kilometer distant (Wendt 1964). Platform mounds received even greater elaboration at Aspero in the Supe Valley. At this location, there is a complex of at least six artificial mounds. Most were terraced, and all have traces of stone facings and/or masonry summit structures (Moseley and Willey 1972).

On the basis of sites such as El Paraíso, it is evident that the maritime peoples were fully capable of undertaking relatively extensive building projects. Río Seco and Aspero make it obvious that there was a sufficient level of organization for people to engage in corporate labor ventures that were esoteric in nature, and did not directly confer material benefits upon the builders themselves.

The nature of social organization among the early populations of coastal Peru is presently unknown. However, it is apparent that marine resources were supporting substantial numbers of people, many of whom resided in large nucleated communities. Furthermore, at these communities organization was such that mass labor forces could be mobilized to modify the natural terrain according to preconceived plans for either practical or esoteric purposes. Given this background, it is not surprising that large-scale construction of canals could proceed rapidly. In other words, the use and development of irrigation technology was not restrained by the necessity to formulate the social principles needed to integrate and mobilize corporate work parties. Both the requisite labor force and the organization principles were pre-existent.

In summary, if the archaeological record implies an accelerated development of canal irrigation during the initial stages of use, it also implies that the early coastal societies were fully capable of having implemented this development. Prior experience with both plant cultivation and corporate labor construction conferred a degree of cultural preadaptation for the utilization of this water management technique, and such preadaptation facilitated the rapid evolution of canal irrigation.

## CONCLUSION

It is worthwhile to examine briefly two phenomena related to early irrigation and intensive farming: the introduction of agricultural staples to the coast and the role of population pressure upon marine resources.

The introduction of cultigens suitable for food staples was a significant factor in the process of raising the level of agricultural productivity and making farming more economically competitive than fishing. Potential staples were not indigenous to the coast, but had to be introduced via diffusion from outside sources. Therefore, the inception of intensive farming was closely related to the building up of a repertoire of productive plants, as well as to the development of water management techniques. Even with canal irrigation of the desert, an absence of staples would curtail an effective rise in agricultural output. On the other hand, the growing of staples in the context of floodwater farming would have had limited potential because land suitable for this type of cultivation was not abundant.

The nature of the feedback between the introduction of agricultural staples and the development of irrigation is not well understood. Maize appeared at a limited number of coastal sites shortly before the advent of inland farming. Yet, even with the inception of irrigation the use of this plant diffused relatively slowly, although it ultimately became very important as a food source. On the other hand, at the time irrigation is postulated to have begun in the Chillón Valley, there was the concomitant appearance of a number of productive crops such as sweet potatoes, peanuts, and certain legumes, as well as a variety of different fruits. Many of these crops also appeared in a comparable early context in coastal valleys further to the south. These lines of evidence point to an intimate relation between the inception of irrigation and the introduction of domesticated staples, but they do not imply that one necessarily caused the other. Rather, both combined to create a significant increase in the potential level of agricultural output.

The demographic situation at certain early coastal communities was probably an important variable affecting the initial phases of intensive agriculture. A recent review of the larger maritime settlements shows that sites located near valley mouths tend to be somewhat larger than isolated desert settlements. The size differences are apparent, but not particularly marked, with the exception of El Paraíso at the mouth of the Río Chillón (Moseley and Willey 1972). In this review, I hypothesized that intensive farming evolved out of a feedback process involving demographic pressures that had built up at certain valley mouth, maritime settlements (Moseley 1972). Such sites could serve as bases for both fishing and farming. Although floodwater farming was not particularly productive, it did allow the population size of certain sites to rise above the level that could be maintained by the locally available marine food supply. Once the population grew beyond the maritime support capacity – even if the growth was limited – there had to follow an economic commitment to more than casual farming if the extended settlement size was to be maintained. In other words, the hypothesis postulates a buildup of demographic pressures at certain sites as fostering a certain degree of economic dependence upon farming. If this argument is correct, the demographic situation at some valley mouth settlements was an important factor in the rise of intensive agriculture.

If the archaeological record from the Peruvian coast has been correctly read, the advent of early intensive cultivation of the desert involved at least three major variables: 1) a significant degree of cultural preadaptation to canal irrigation; 2) the development of a productive repertoire of domesticated staples; and 3) a demographically precipitated commitment to more than casual farming at certain maritime settlements. Some time around 1750 B.C., these variables coalesced to bring about the rapid formulation and development of intensive agricultural exploitation of desert lands.

## REFERENCES


BIRD, Junius B.

1948 Preceramic Cultures in Chicama and Virú. *Memoirs of the Society for American Archaeology*, No. 4, pp. 21-28.

ENGEL, Frederic

1958 Sites et établissements sans céramique de la côte péruvienne. *Journal de la Société des Américanistes* 46:67-155.

1963 A Preceramic Settlement on the Central Coast of Peru: Asia, Unit 1. *Transactions of the American Philosophical Society* Vol. 53, part 3.

1966 Le complexe précéramique d'El Paraiso (Pérou). *Journal de la Société des Américanistes* 55:43-96.

GAYTON, A. H.

1967 Textiles from Hacha, Peru. *Ñawpa Pacha* 5:1-14.

LANNING, Edward P.

1963 A Pre-agricultural Occupation on the Central Coast of Peru. *American Antiquity* 28:360-71.

1967 *Peru Before the Incas.* Englewood Cliffs: Prentice-Hall.

MOSELEY, M. Edward
1968 Changing Subsistence Patterns: Late Preceramic Archaeology of the Central Peruvian Coast. Ph.D. dissertation, Harvard University.
1972 Subsistence and Demography: An Example of Interaction from Prehistoric Peru. *Southwestern Journal of Anthropology*. 28:25-49.

MOSELEY, M. Edward and Gordon R. Willey
1972 The Preceramic Status of Aspero, Peru. MS.

MURPHY, R. C.
1923 The Oceanography of the Peruvian Littoral with Reference to the Abundance and Distribution of Marine Life. *The Geographical Review* 13:64-85.

PATTERSON, Thomas C. and Edward P. Lanning
1964 Changing Settlement Patterns on the Central Peruvian Coast. *Ñawpa'Pacha* 2:113-23.

PATTERSON, Thomas C. and M. Edward Moseley
1968 Late Preceramic and Early Ceramic Cultures of the Central Coast of Peru. *Ñawpa Pacha* 6:115-33.

PICKERSGILL, Barbara
1969 The Archaeological Record of Chili Peppers (Capsicum spp.) and the Sequence of Plant Domestication in Peru. *American Antiquity* 34:54-61.

SVERDRUP, Harald Urik, Martin W. Johnson, and Richard H. Fleming
1946 *The Oceans: Their Physics, Chemistry and General Biology*. New York: Prentice-Hall.

WENDT, W. E.
1964 Die präkeramische seidlung am Rio Seco, Peru. *Baessler Archiv* 11:225-75.

CHAPTER 8

# IRRIGATION AND SETTLEMENT PATTERN
## Preliminary Research Results from the North Coast of Peru

Ian Farrington
*Centre for Latin-American Studies, University of Liverpool*

The Peruvian coast is an almost totally arid desert, yet in pre-Hispanic times it was the locus of development of advanced preindustrial cultures based primarily upon agriculture. Cultivation there is only possible with an adequate supply of water, which may be obtained from the flooding of the exotic streams or more securely from canal systems and underground resources. The era of incipient agriculture, the Late Preceramic and Initial Periods (2500 B.C. to 900 B.C.),[1] is thought to have utilized floodwater farming because of the minor nature and scarcity of the domesticates in the archaeological record (Engel 1966; Moseley 1972). This technique has been regarded elsewhere as a precursor of irrigation because it involves a certain amount of control of the collection, transfer, and distribution of water resources (Farrington 1971). After these periods, the nature of the domesticates and the overwhelming importance of plant foods in the diet imply canal irrigation, of which vestiges can be traced from the Early Intermediate Period (ca. 300 B.C. to ca. 900 B.C.) until the Spanish Conquest in 1532. Thus the history of agriculture on the coast of Peru is the history of irrigation. Water management is the dominant agricultural technique throughout the sequence and, for the most part, the canal system delimits the area under cultivation.

The importance of irrigation in Peruvian prehistory has been recognized by many authorities, but specific studies of it are lacking, with the result that relationships between it and various aspects of society cannot be elucidated. Basically, the published works can be grouped into four classes:

(a) valley by valley descriptive and bibliographic surveys which offer little analysis (Regal 1945, 1970; Tord 1969).

(b) studies of specific valleys by the interpolation of modern hydrological data into the past, but with little consideration of the hydraulics and importance of the ancient systems (Rodriguez Suy Suy 1970).

(c) studies in which irrigation is invoked to be a prime causal factor in the development of social stratification and urbanism, and the nebulous relationships between irrigation and its bureaucratic management are discussed (Kosok 1965; Lanning 1967; Steward 1955; Wittfogel 1957). These are based not only on an imperfect understanding of irrigation techniques and operation, but also on the nature of prehistoric Peruvian society.

(d) works on settlement patterns which did not study water management *per se*, but used it as an aid to interpret the data. Willey's (1953) study of the Virú Valley had as its primary concern the interpretation and dating of sites and subsequent settlement and "community pattern"[2] analysis, and from the latter he was able to make inferences about social organization. He realized that there was a spatial relationship between sites and remnant irrigation features, but mere proximity of one to the other dated them contemporaneously. He discovered the spatial relationship without study of the irrigation networks themselves or of the direct relationship between sites and canals and fields, i.e. their location within, adjacent, beyond, or above agricultural land, or adjacent to or cut by canals. Consideration of these factors changes the dating, but not the underlying assumption that there is a relationship between irrigation and settlement pattern.

The two methodological statements – that there is a relationship a) between irrigation and settlement pattern; and b) between settlement pattern and social organization – are mutually exclusive. It is not possible to state that irrigation implies social organization or vice versa because the first is a relationship between two actual spatial distributions and is reversible, whereas the second is a spatial distribution which manifests social pattern. It is a social relationship and is irreversible, i.e. social organization or community pattern does not imply settlement pattern.

It is a contention of this paper that a study of irrigation agriculture on the north coast of Peru can only be carried out by the analysis of the relationships between it and the physical and cultural environments. In the first place, there is a technological relationship between irrigation or land use and ecological zones, and secondly, there is the previously mentioned correlation between settlement pattern and water management. It is, then, settlement form, location, and pattern which, in the absence of documentary sources, gives some impression of social organization.

In order to examine these settlement patterns, it is necessary to outline the major conclusions of the Virú report (Willey 1953), because it is on the basis of these patterns that research was organized. First of all, it was noted that there was a concentration of prehistoric sites on the valley margins around the limits of modern cultivation, which was explained to be the result of site destruction in the flood plain by recent farming. Secondly, it was stated from settlement evidence that canal irrigation began in the Puerto Moorin Period (early phase of the Early Intermediate, 400 B.C. to 0), although the first physical evidence is from the next phase, the Gallinazo (0 to 800 A.D. This phase probably ended around 300 A.D.). During the latter, irrigation was highly developed and thought to be at its maximum pre-Hispanic extent. Irrigation continued to be important throughout the remainder of the sequence with some reductions in area, but a new technique, sunken field agriculture (*pukio*), slightly increased the area under cultivation when it was introduced in the Tomaval (Early Chimú, 1100 to 1300 A.D.). Willey further remarked upon a depopulation evident in the settlement pattern in La Plata (Imperial Chimú, 1300 to 1460/1470) and Estero (Chimú-Inca, 1460/1470 to 1532) times. He attributed the first to the centralization policy of the Chimú Empire and its focussing upon Chan Chan, and the second to the Inca Conquest. Finally, Willey discussed the presence of mounds with a high concentration of salt in the lower valley flood plain. These, he explained, were accumulations of salitre which had been scraped off the fields to permit agriculture. In other words, he recognized an ecological problem and solution in the use of these bottom lands.

The valleys of the north coast may be divided into a variety of ecological zones, but for the purposes of this analysis it is sufficient only to consider three (Fig. 8.1 illustrates these for the Moche Valley):

(i) the area immediately adjacent to the river, which is characterized by a high water table, springs, heavy poorly drained soils, and thick vegetation. A similar zone exists on the littoral behind old beach lines. In both of these salinization of the soil is a problem.

(ii) the upper terrace lands of the flood plain, which have easily drained fertile soils and a less dense vegetation cover.

(iii) the valley sides, which are devoid of vegetation and have only a weakly developed soil or none at all. These comprise sand deserts and desert pavements.

All valleys, too, have similar water regimes – a surfeit during the months of January, February, and March, when there is rain in the mountains, and a period of diminished flow from July to October.

In pre-Hispanic times, agricultural techniques were adapted to exploit these zones and water regime. Water management was such that it may be described as sophisticated floodwater canal irrigation involving simple methods of collection, transfer, and distribution of water, but with virtually no storage. Most of the canals carried water only during the summer months which thus enabled the production of only one crop per year. This conclusion has been confirmed by recent analysis of a midden in the Moche Valley, which revealed an annual cycle of food remains and included only one peak of agricultural products (Griffis 1971). Furthermore, the inventory of agricultural tools known on the coast did not include a heavy clod cutter such as the *chaquitaclla* (foot-plough). Nor were there any techniques of drainage or water table utilization practised before the last two periods, Imperial Chimú and Chimú-Inca. These last two points are important in understanding the settlement pattern from the Initial Period until the Chimú-Inca.

Recent work in the Moche Valley has confirmed that the majority of ancient sites, such as public structures, cemeteries, and those of a domestic nature, do lie on the valley margins, and did so in all periods. Agricultural land was always at a premium and, although settlement had to be constructed nearby to enable efficient operation, it was located on land not required by the agriculture of the time. Sites were often above and adjacent to the uppermost canal or on low mounds and hillsides within the cultivation area. The only structures actually within the fields were minor shelters and rooms. Thus, by plotting the distribution of major sites of a particular period, the area under cultivation at that time can be estimated. The concentration on the valley flanks through time is further emphasized by the virtual lack of occupation on the lower, wetter flood plain (zone i) after the phases of incipient agriculture. The reasons for this dichotomy are evident in the technological level of agriculture. The lands of the valley sides and upper flood plain terraces (zones iii and ii) have light, free-draining soils which are easily cleared of vegetation and can be prepared for sowing by use of a simple digging stick. They have to be irrigated little but often (Vidalón and Guzmán 1965), which suits admirably the type of water system which was in use and, therefore, with minimal effort could produce one crop a year. The bottom lands remain empty because there is too much water; successful cultivation would require drainage and spring control. It was not until late in the north coast sequence that techniques became available and/or there was a demand to open up zone (i). Steep hillsides in the lower valleys, such as Cerro Orejas in the Moche, were terraced during Moche times (later phases of the Early Intermediate Period, 300 A.D. to 1100 A.D.) to avoid the heavy bottom lands adjacent to the river. There is an Early Intermediate Period *huaca* (mound) in the flats off the foot of this mountain, on land which was probably unused. The terraced hillside provided easier and more economic land to cultivate.

In the Moche and Virú Valleys, the pattern of land-use expansion and exploitation followed these principles and can be traced from the beginning of the reliance upon agricultural products in the late Initial Period and Early Horizon (1200 B.C. to 300 B.C.). Prior to this time, there is evidence of small-scale floodwater farming in coastal and flood plain situations. Irrigation agriculture began as an upper valley phenomenon, with short canals leading to easily cultivated lands. The area under cultivation in the Early Horizon was small, but as far as can be gathered from settlement pattern analysis, it comprised a number of discrete sections on different canals. These sections were mainly in the narrow parts of the valleys, but it is presumed that in the Moche there was one on the lower valley terraces (zone ii) in an area which is now agricultural land. This area is downhill from Caña Huaca, a late Early Horizon mound.

The canal pattern was merely extended during each successive period as more land was required for cultivation (Fig. 8.2). New lands were opened up beyond and even over older sites, for cultivable land and water were never major resource problems in these valleys. Gallinazo was merely one phase in this development and was not, as Willey suggested, the period of maximum canal extent in either the Virú or Moche Valleys. By late Moche times (600 A.D. to 900 A.D.) and the Moche-Chimú transition (900 A.D. to 1100 A.D.), most of the valley side lands had been brought into cultivation by the extension of existing canals. In the Moche Valley new land was sown on the low divide between it and the Rio Seco to the north. This necessitated the construction of a large aqueduct to take water across cultivated land to the new area, the Pampa La Esperanza (Farrington 1972). The colonization of this area is documented by a series of small mounds and a winged structure of probably administrative function located on the uphill side of the canal which leads from the aqueduct, i.e. outside the cultivated area. Within this area, however, are a series of small one- or two-roomed field structures of similar late Moche date (Fig. 8.3). The dominant settlement pattern is still that of sites located outside but proximal to cultivated land. On the basis of the analysis, there is no evidence of direct control of the irrigation system by any particular site. In neither this nor any other period were sites located at major intakes or bifurcations along the canal network. There were large ceremonial and domestic centers in each valley from Nepeña to Lambayeque during this time, such as Huaca del Sol and Galindo in the Moche, Mocollope in Chicama and Huancaco in Virú. These are believed to be the centers of valley-wide or even multi-valley states.*

The Early Chimú Period (Tomaval in the Virú Valley) saw a continuation of the previous pattern, but with some decline in importance of the large ceremonial centers. This period was followed by the Imperial Chimú Period, when the whole of the settlement pattern, not only of the Moche but also of the Virú and Chicama, became dominated by the urban center of Chan Chan. There was a large nucleated population around this ceremonial, storage, and redistributive center. On the coastal pampa there were some canal extensions and new land schemes, particularly the opening up of the northern section of the Pampa La Esperanza and the Pampa El Milagro (Fig. 8.2). In both of these areas small administrative buildings were constructed within the fields. The erection of these buildings indicates an important

*M.E. Moseley 1972: personal communication.

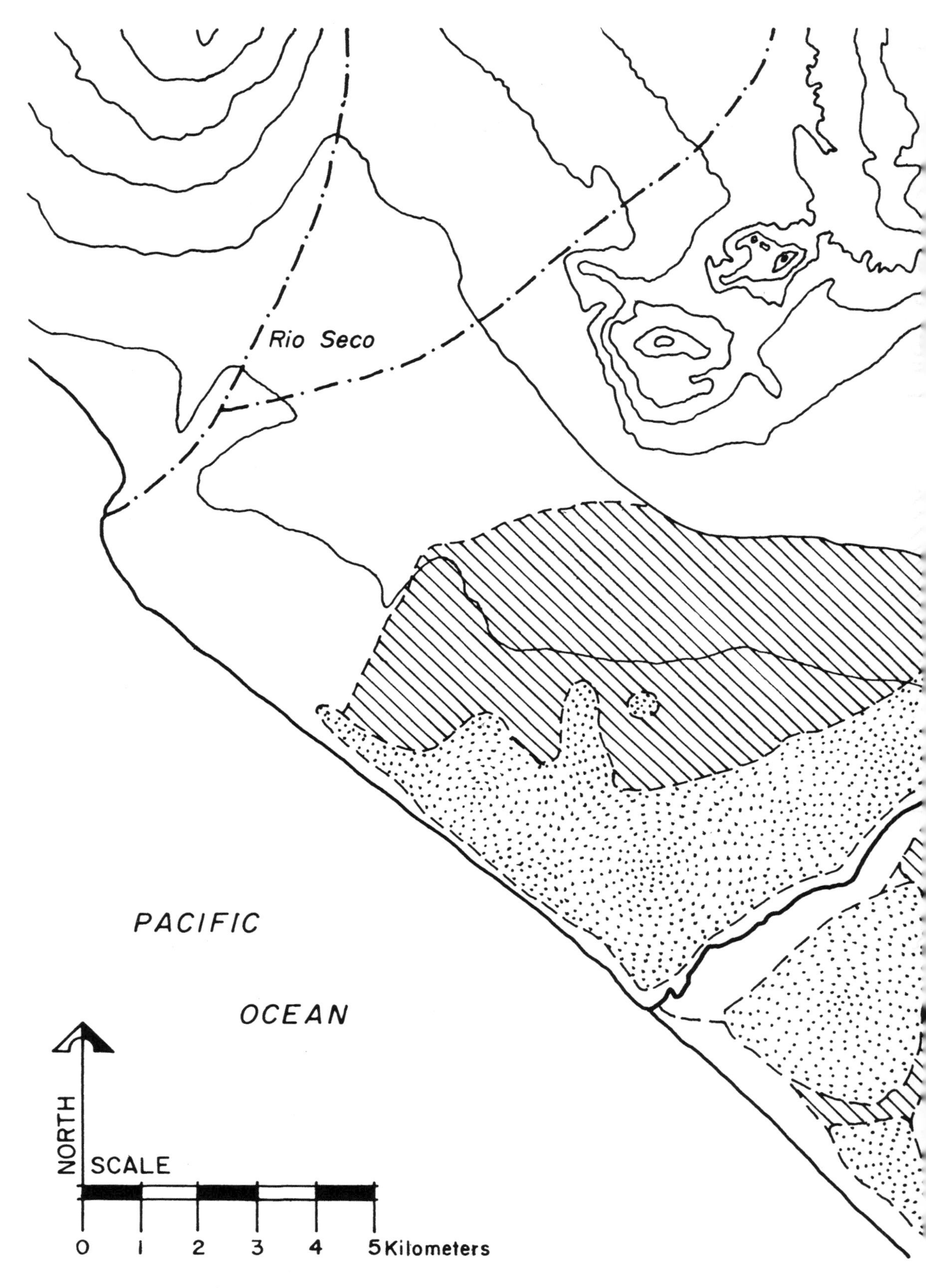


Fig. 8.1 Map: major ecological zones of the Moche Valley.

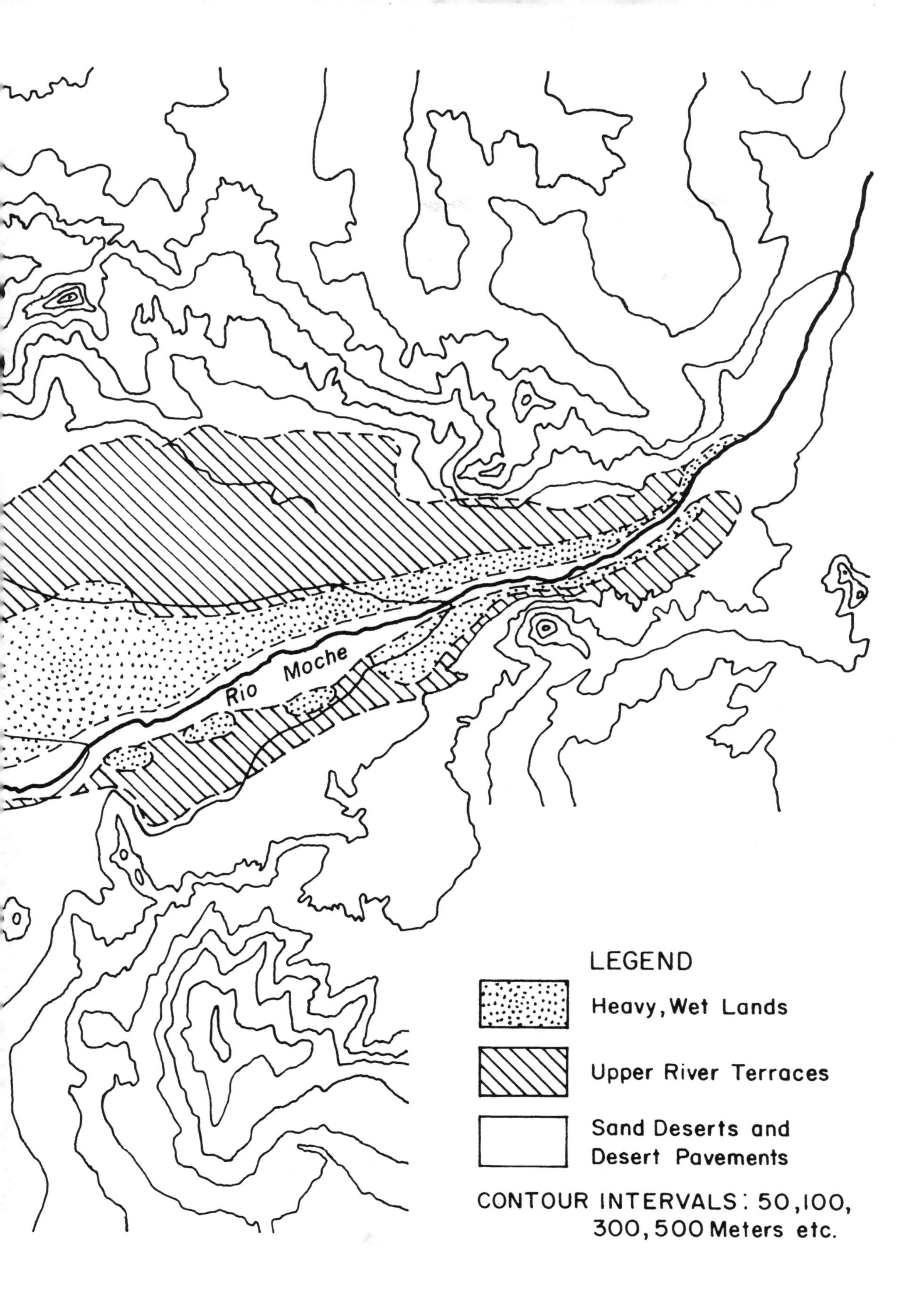
Rio Moche
LEGEND
Heavy, Wet Lands
Upper River Terraces
Sand Deserts and Desert Pavements
CONTOUR INTERVALS: 50,100, 300, 500 Meters etc.

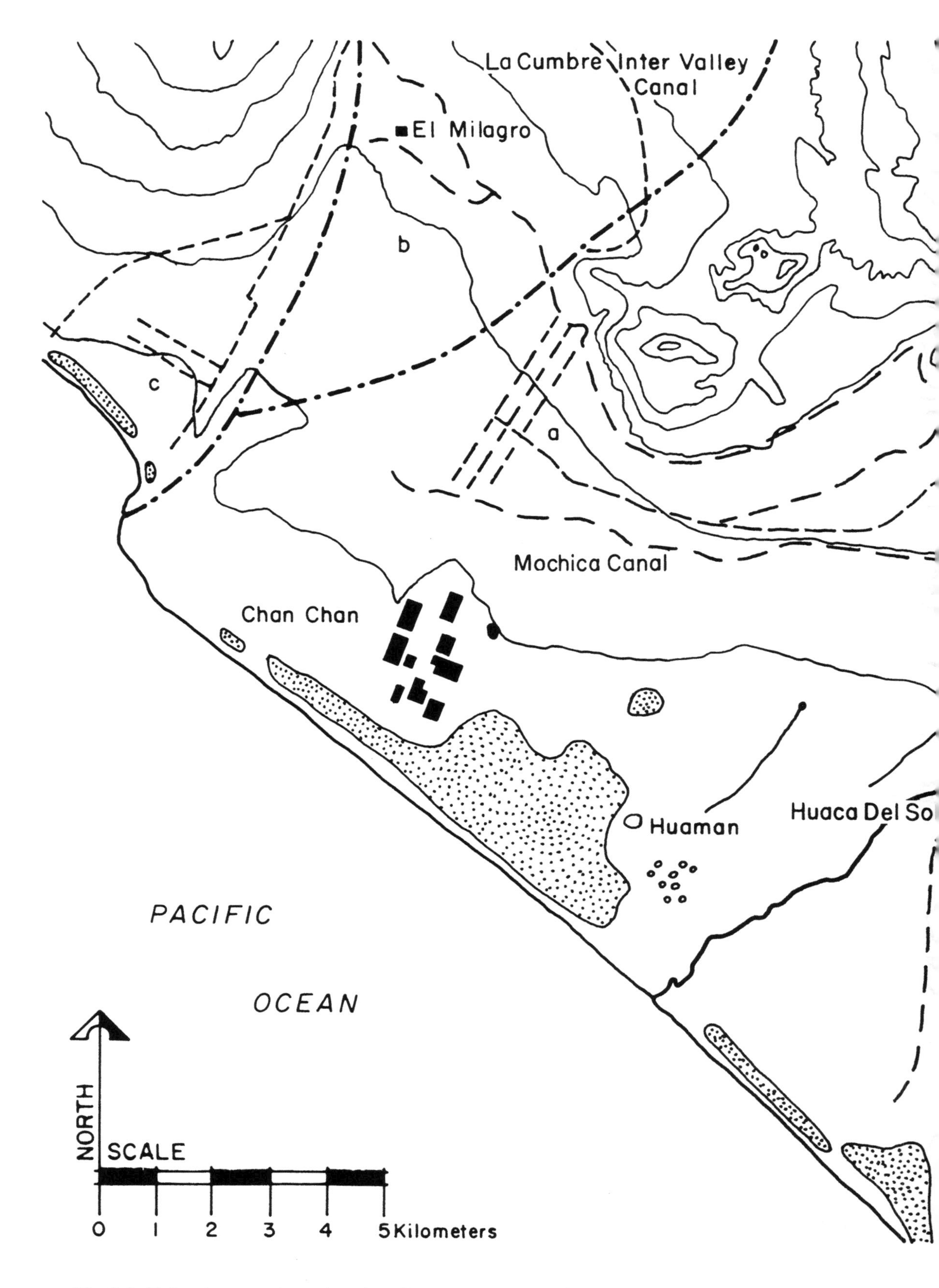


Fig. 8.2 Major sites, canals, and sunken fields in the Moche Valley.

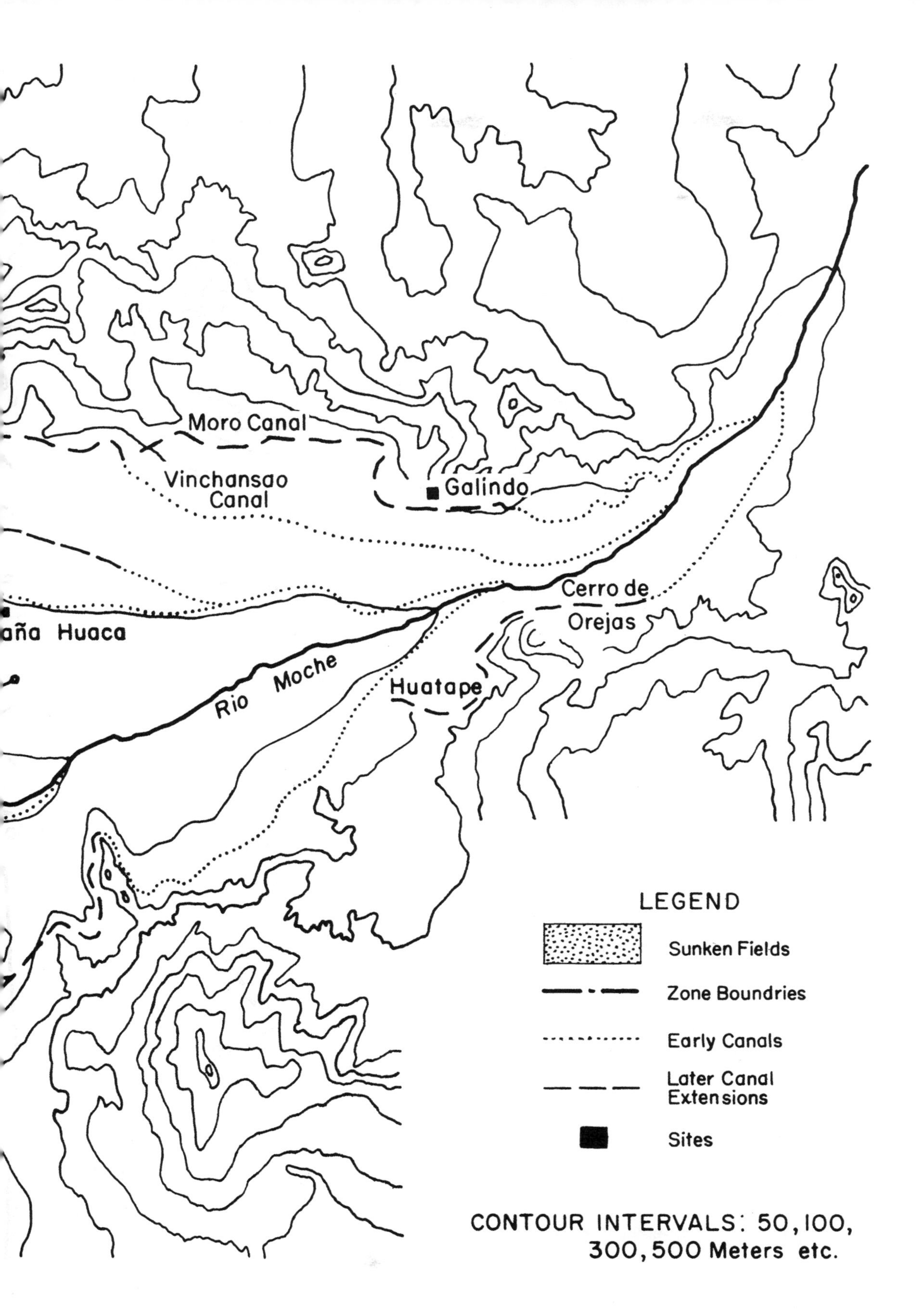

Moro Canal
Vinchansao
Canal
Galindo
aña Huaca
Rio Moche
Cerro de
Orejas
Huatape
LEGEND
Sunken Fields
Zone Boundries
Early Canals
Later Canal
Extensions
Sites
CONTOUR INTERVALS: 50, 100,
300, 500 Meters etc.

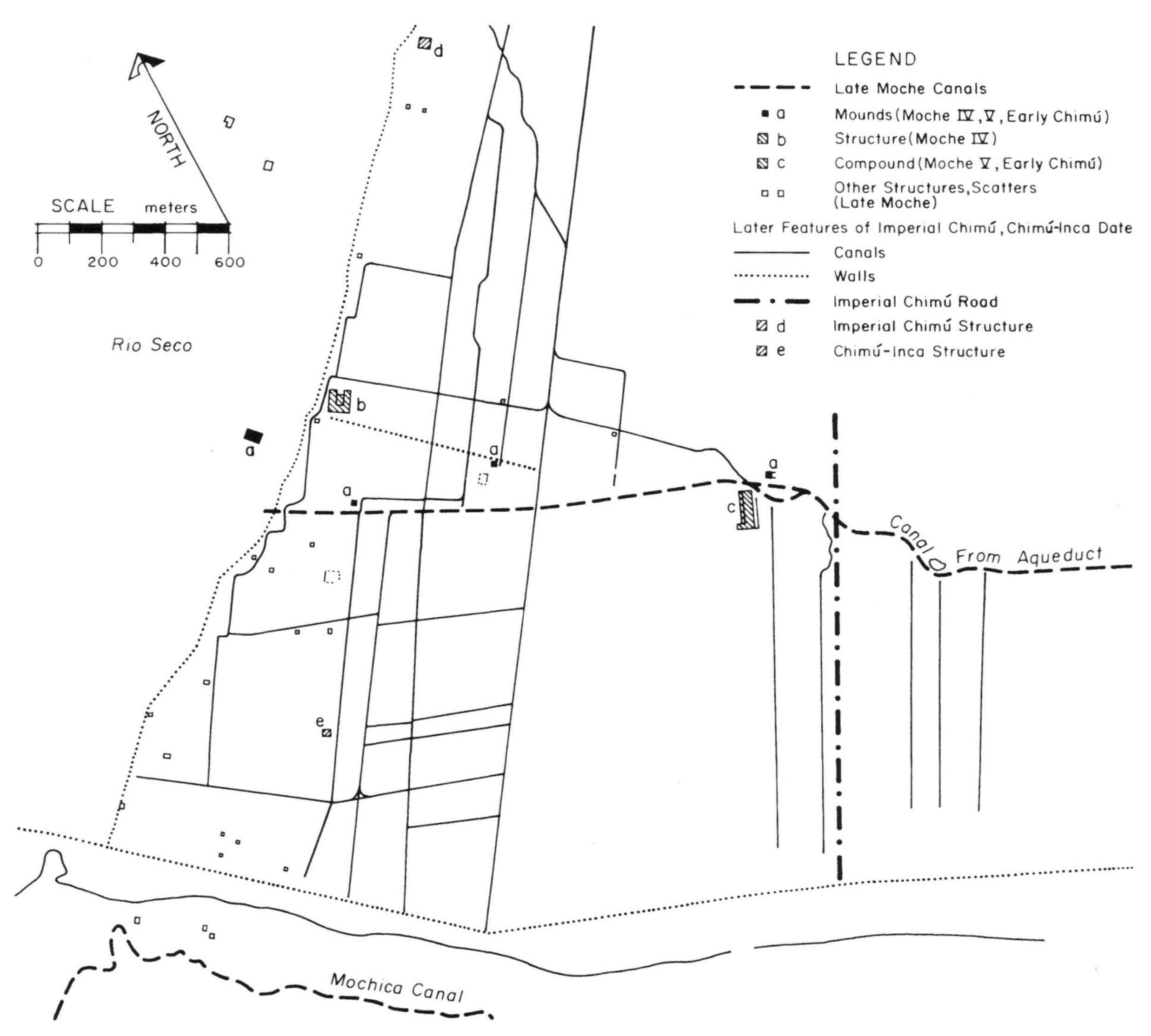


Fig. 8.3 Late Moche colonization of the lower Pampa la Esperanza.

change in settlement pattern in the valley, one of land values, which is obviously related to the function of the building. It is suggested that these were new state lands operated by the central government via these structures. Presumably the other lands were worked as before, but probably from residences in the urban area. In Virú there were no major sites and a marked decrease in the number of sites; Willey attributed this phenomenon to the centralization policy of the Chimú Empire. The sites which were occupied were in the traditional situation, although there are indications in the Virú Report that the coastal lagoon area was being exploited at this time. It is known from the Moche Valley that groundwater sources were first employed for urban purposes at this time (Day 1970), but survey of sunken field areas in these valleys has placed their initiation in the Chimú-Inca Period.

After the conquest of the Chimú Empire by the Inca in the 1460s, there was no sizable depopulation as presumed by Willey. Many of the sites listed by him as Early Chimú are, in fact Chimú-Inca; such redating is the result of a detailed redefinition of pottery styles. In fact, there was a massive expansion in land under cultivation; canals were extended and new schemes undertaken. The Pampa El Milagro was significantly enlarged and altered, and laid out with a regular field and furrow pattern. The Pampa Cerro de la Virgen was opened up with very regular, rectangular fields with straight furrows. Initially, water for both of these areas was carried out of the Moche by the Acequia Vichansao, which resulted in the production of only one crop per annum (Griffis 1971). During the Inca Period, however, the La Cumbre or Intervalley Canal was constructed to take water from the larger Chicama River to the Vichansao at a point to the north of Chan Chan. The Chicama River has a much greater flow for a much longer period than the Moche, and it is presumed that its water would have been used to supplement irrigation of the Pampa El Milagro and Pampa Cerro de la Virgen, and also to increase the number of croppings on these. At the junction of the Intervalley Canal with the Vichansao, a reservoir was constructed to regulate the timing of irrigation, but from archaeological evidence it appears that it was never used. Surviving areas of fields of this period tend to have a very regular layout, and coupled with the construction of this major canal, imply some form of organized planning authority similar to that suggested for the Inca heartland (Moore 1958).

Settlement associated with the new field areas comprises basically two types: simple field shelters of U- or L-shape found in pairs throughout the fields (Fig. 8.4*d*) or larger, more complex, baffled, stone structures with wings to form an entry court (Fig. 8.4*a* and *c*). There is a hierarchy of size and complexity of the latter type, with the more elaborate having as their focus niched rooms with attendant storerooms (Fig. 8.4*a* and *b*). This form is identical in layout and, presumably, in function to the storage areas of both Imperial and Inca Chan Chan (Andrews 1972). The largest of these, El Milagro de San José in the Río Seco drainage, is the dominant structure in a small complex of such buildings. It is surrounded by furrows and was obviously concerned with the administration of them. Most of these structures occur in groups of two or more; often each component is of a different order, or is associated with a group of shelters. None has a large domestic occupation. The domestic sites associated with these new lands are in traditional locations unwanted for agriculture.

During the Chimú-Inca Period, new techniques were introduced or there was a demand to employ known techniques to colonize the wetter, heavier lands of the flood plain and littoral (zone i). This colonization probably involved the regulation of spring flow by enlarging the 'weep' area, and canalization, which had the added effect of draining the land, making it possible to work. The foot-plough is known to have been used in the Highlands by the Inca, and there is a strong possibility that it was introduced to the coast by them.[3] Sunken fields were also constructed in the backswamps behind old beach lines and in other inland areas of high water table, such as the Ciudad Universitaria in the Moche Valley. In the Virú, Moche, and Chicama Valleys there is no evidence to support dating sunken field construction to the Early Chimú or Imperial Chimú Periods. All sherds collected from associated settlements and from excavations of sunken field banks imply Chimú-Inca construction. Settlement in these areas of poorly drained soils tended to be concentrated on artificial mounds or ridges. One important architectural site in Virú is situated within these lands, and has a form and niched room similar to the one previously described. This is V-124 (Willey 1953:324).

In contrast, Chimú-Inca settlement in the valleys proper, i.e. away from the new lands, continued the traditional pattern. Sites were all adjacent to, above, or beyond the cultivated land; even the form of settlement was different. There continued to be much domestic terracing, but there were also seven larger structures of sub-rectangular form with little distinguishable internal subdivision, containing a large amount of domestic refuse (Fig. 8.4*e* and *f*). They were all associated with areas of poorer domestic architecture and cemeteries.

This dichotomy of settlement form and distribution is a result of the breakdown of the rigorous centralized government of Chan Chan and the implantation of Inca policy. Chan Chan remained a large storage and redistributive center with a considerable population, but deliberate decentralization of part of its authority did occur. The settlement in the valleys represents the return by the local *curacas* (lords) to their homes with jurisdiction over their own lands (although tribute would still have had to be paid to the Inca). For example, in the area known as Huatape in the Moche Valley, there is a large Chimú-Inca occupation of large structures, smaller domestic ones, and cemeteries in an area first colonized agriculturally in Early Chimú times and displaying no significant Imperial Chimú evidence. This occupation confirms the arguments put forward by Rowe (1948) that the Inca preserved the native system of landholding providing the *curaca* did not oppose them. On the Central Coast, there is documentary evidence of *curaca* opposition to the Inca and the lands of the dissenters being handed over to *mitimaes*, or loyal settler, groups.* The areas of land colonization which are characterized by winged and baffled structures are believed to have been operated by the Inca state itself

---

*Rostworowski de Diez Canseco 1972: personal communication.

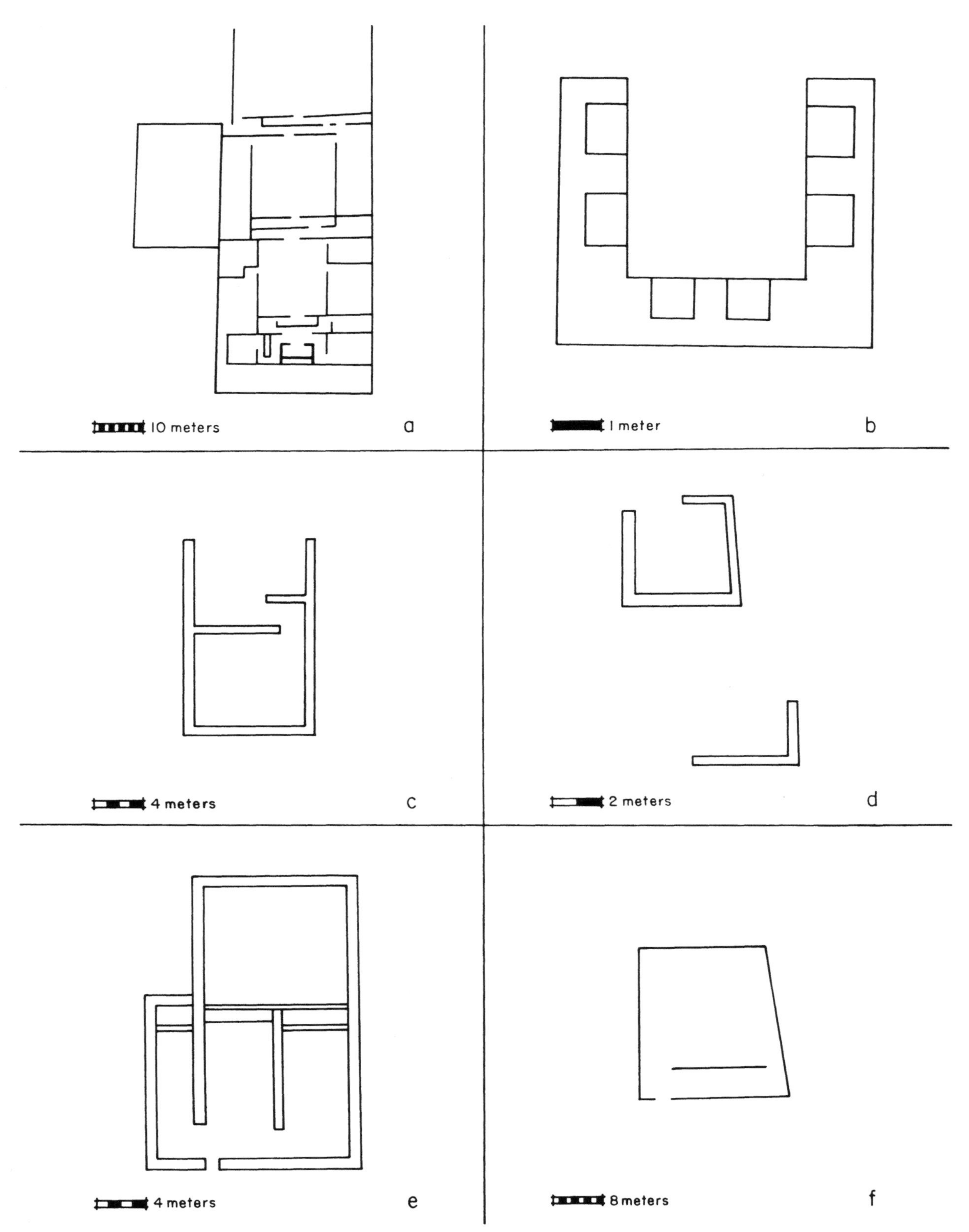

Fig. 8.4 Chimú-Inca structures.

to supply the needs of the church and state. It is known that the Inca did open up new lands for these purposes in the Highlands, and there is no reason to dismiss the practice of the same policy on the North Coast. In order to farm the new lands, many forms of labour were employed: *yanacona,* a kind of royal retainer group; *mit'a*, or corvée type of system; and *mitimaes.* The latter group consisted of loyal agriculturalists who were settled in newly conquered areas to prevent insurrection, and were given unused land to cultivate for themselves. The new lands were colonized by both state and *mitimaes* policies in the Highlands, and consisted of not only fertile flat areas and terraced hillsides, but also poorly drained areas where springs were enlarged and streams canalized (Rostworowski de Diez Canseco 1962). In Virú the poorly drained zone was administered by the state from the previously mentioned V-124, whereas on the north side of the Moche River a different pattern emerged. In that area there are a number of Chimú-Inca occupation mounds above the flats. There is also a Spanish *reducción*[4] settlement called Huaman, a Quechua name. This place-name and surface evidence suggest a strong possibility of *mitimaes* from the Highlands being brought into the Moche Valley for political reasons and farming an area, hitherto unwanted, with their techniques of drainage, spring 'weep' improvement and canalization, and the foot-plough. The old lands in the valley and the new lands of zone (i) continued the traditional settlement pattern, although in the latter occupation mounds were artificially constructed. The only divergence from the traditional pattern was in the case of new state lands with administrative buildings within the fields and not at strategic locations along the canal network. In Chimú-Inca administration and, from settlement evidence, that of other periods as well, irrigation was never of paramount importance.

In conclusion, it may be said that there is a very strong spatial relationship between irrigation agriculture and settlement pattern on the North Peruvian Coast, but that this relationship is totally independent of social organization. The latter is interwoven with settlement pattern alone, its only spatial manifestation.[5]

## NOTES

[1]The dating scheme employed in this paper is an amalgamation of Rowe's scheme (1962) with specific north coast studies by Larco Hoyle (1948) and Ford (1949) and the recent refining of the latter by C. J. Mackey, U. Klymyshyn and G. Conrad (all members of the Peabody Museum Moche – Chan Chan Project). The absolute dates cannot be taken as wholly accurate because of the lack of radiocarbon assays.

[2]Willey (1953:371-89) used the term "community pattern" as an analytical tool to discern the pattern and arrangement of society from spatial characteristics.

[3]R. A. Donkin in a recent paper (1970:515) illustrates a complex pot which comprises a *chaquitaclla* attached to an aryballoid vessel. The aryballus is a diagnostic Inca form, although this one is black and has been given Chimú area provenience. His work uses ethnographic evidence to describe its utilization and distribution solely in the Highlands since the Conquest. He gives as one reason for the disappearance of the tool from much of Peru as: "the widespread abandonment of the heavy bottom lands."

[4]*Reducción* is the term given to the early Spanish colonial policy of congregating the Indians into nucleated settlements in order to facilitate administration.

[5]The author wishes to acknowledge the support of the Department of Education and Science, London (research grant), the Central Research Fund of the University of London (grant for aerial photographs) and the Peabody Museum Moche – Chan Chan Project in Trujillo, particularly the following members of it: M. E. Moseley (Director), C. J. Mackey, G. Conrad, U. Klymyshyn, and L. Watanabe.

## REFERENCES

ANDREWS, A.
1972 Architecture (of U-shaped, niched Structures in Chan Chan), MS.

DAY, K.
1970 Walk-in Wells and Water Management at Chan Chan, Peru. Paper read at the 39th Congreso Internacional de Americanistas, Lima.

DONKIN, R. A.
1970 Pre-Columbian Field Implements and Their Distribution in the Highlands of Middle and South America. *Anthropos,* 65, (3/4), 505-29.

ENGEL, F.
1966 Le Complexe Précéramique d'El Paraiso (Pérou). *Journal de la Société des Americanistes,* 55, 43-96.

FARRINGTON, I. S.
1971 Towards a Classification of Primitive Irrigation Techniques with Special Reference to the Pre-Hispanic Americas. MS.
1972 El Acueducto de Mampuesto en el Valle de Moche. Paper read at El Primer Congreso de Hombre y Cultura Andina, Lima.

FORD, J. A.
1949 Cultural Dating of Pre-Hispanic Sites in the Virú Valley, Peru. *Anthro. Papers Amer. Museum Nat. History,* 43, (1).

GRIFFIS, S.
1971 Excavation and Analysis of Midden Material from Cerro La Virgen, Moche Valley, Peru. MS, Unpub. B.A. thesis, Dept. of Anthropology, Harvard Univ.

KOSOK, Paul
1965 *Life, Land, and Water in Ancient Peru.* New York: Long Island Univ. Press.

LANNING, E. P.
1967 *Peru Before the Incas.* Englewood Cliffs: Prentice-Hall.

LARCO HOYLE, R.
1948 *Cronología Arqueológica del Norte del Perú.* Buenos Aires: Sociedad Geografia Americana.

MOORE, Sally Falk
1958 *Power and Property in Inca Peru.* New York: Columbia Univ. Press.

REGAL, A.
1945 La Política Hidráulica del Imperio Incaico. *Revista de la Univ. Católica del Perú,* 13, (2-3), 75-110.
1970 *Los Trabajos Hidráulicos del Inca en el Antiguo Perú,* Lima.

RODRIGUEZ SUY SUY, V. A.
1970 Irrigación Prehistórica en el Valle de Moche. Paper read at the 39th Congreso Internacional de Americanistas, Lima.

ROSTWOROWSKI de DIEZ CANSECO, M.
1962 Nuevos Datos Sobre Tenencia de Tierras Reales en el Incario. *Revista del Museo Nacional* (Lima), 21, 130-59.

ROWE, J. H.
1948 The Kingdom of Chimor. *Acta Americana,* 6, (1), 26-49.
1962 Cultural Unity and Diversification in Peruvian Archaeology. Fifth International Congress of Anthropological and Ethnological Sciences, Phila., 627-31.

STEWARD, J. H. (ed.)
1955 *Irrigation Civilizations: A Comparative Study.* Pan American Union Social Science Monographs 1, Washington D.C.

TORD, J. S.
1969 El Regadío en el Área Andina. *Revista Española de Antropologia Americana,* 4, 113-43.

VIDALÓN, C. and J. Guzmán
1965 *Estudio Agrológico 1e.* Etapa, Proyecto de la Irrigación de Chao, Virú, Moche y Chicama de la Corporación Peruana del Santa, 1, Lima.

WILLEY, G. R.
1953 Prehistoric Settlement Patterns in the Virú Valley, Peru. *Bureau of American Ethnology, Smithsonian Institution Bulletin* 155, Washington, D.C.

WITTFOGEL, K.
1957 *Oriental Despotism: A Comparative Study of Total Power.* New Haven: Yale Univ. Press.

CHAPTER 9

# CONSERVATION AND DIVERSION
## Water-Control Systems in the Anasazi Southwest

R. Gwinn Vivian
*Arizona State Museum*

Until the past few years prehistoric hydraulic systems in the southwestern United States were investigated only sporadically, and it was assumed that water control was of unequal importance among the prehistoric southwestern peoples. The recent work of Haury (1965) and Woodbury (1960, 1961*a*, 1962), however, has confirmed the thesis that irrigation was basic to Hohokam settlement in southern Arizona, and Woodbury's (1961*b*) study of prehistoric agriculture in the Point of Pines area of central Arizona has shown that the practice of water control was equally important among the Mogollon. In this paper I will synthesize information on water control systems of the Anasazi in northern Arizona and New Mexico and attempt to demonstrate that the capture and control of water resources was also a significant cultural development in the northern Puebloan area.

The southern portion of the Colorado Plateau was settled by farming peoples as early as the third century A.D., and palynological evidence suggests that climatic conditions from the third to eighth centuries were conducive to elaboration and expansion of horticulture. Basic similarities in material culture among these peoples contributed to their broad categorization as Anasazi and provided a means for differentiating between them and other major prehistoric culture areas in the Southwest.

Development of Anasazi water-control systems came about almost certainly as a result of changes in climatic conditions. Several geological and palynological studies (Schoenwetter 1970; Schoenwetter and Dittert 1968; Schoenwetter and Eddy 1964; and Harris, Schoenwetter, and Warren 1967) suggest that the southern reaches of the Colorado Plateau underwent a major environmental change from the eighth to twelfth centuries A.D. Although this change varied regionally, essentially it involved a shift from a winter-dominant to a summer-dominant storm pattern. The effects of this shift were shorter and milder winters, a somewhat increased spring drought period, and a "larger number of high energy summer rainstorms with a greater unit runoff rate" (Schoenwetter and Dittert 1968:49).

The Anasazi response to a summer-dominant storm pattern has been discussed by Schoenwetter and Dittert (1968). Adjustments were made through the development of a higher yield, more drought-resistent maize, and more frequent residence changes to locales better suited for horticulture. An equally important response was the development of systems of water control designed to conserve summer moisture.

In the following pages I will consider this latter response in more detail, beginning with a synthesis of the reported data on Anasazi water control systems for the Kayenta, Mesa Verde, northern Rio Grande, and Upper Little Colorado-Zuni areas (Fig. 9.1). Much of the information on these areas was drawn together at the Southwestern Water Control Systems symposium held at the 1970 Pecos Conference in Santa Fe, New Mexico. Drawing on the summarized data, I have proposed a simple classification for Anasazi water control systems into two systems: conservation-type systems and diversion-type systems. This classification involves a consideration of water source, and in reviewing the literature I have distinguished between rainfall, permanent water (perennial streams, springs, etc.), and runoff.

An assessment of the various combinations of control systems and water sources that were attempted by the Anasazi will make it clear that the Anasazi usually opted for systems that would *not* require some form of managerial control of water use. One possible exception to this pattern will be noted

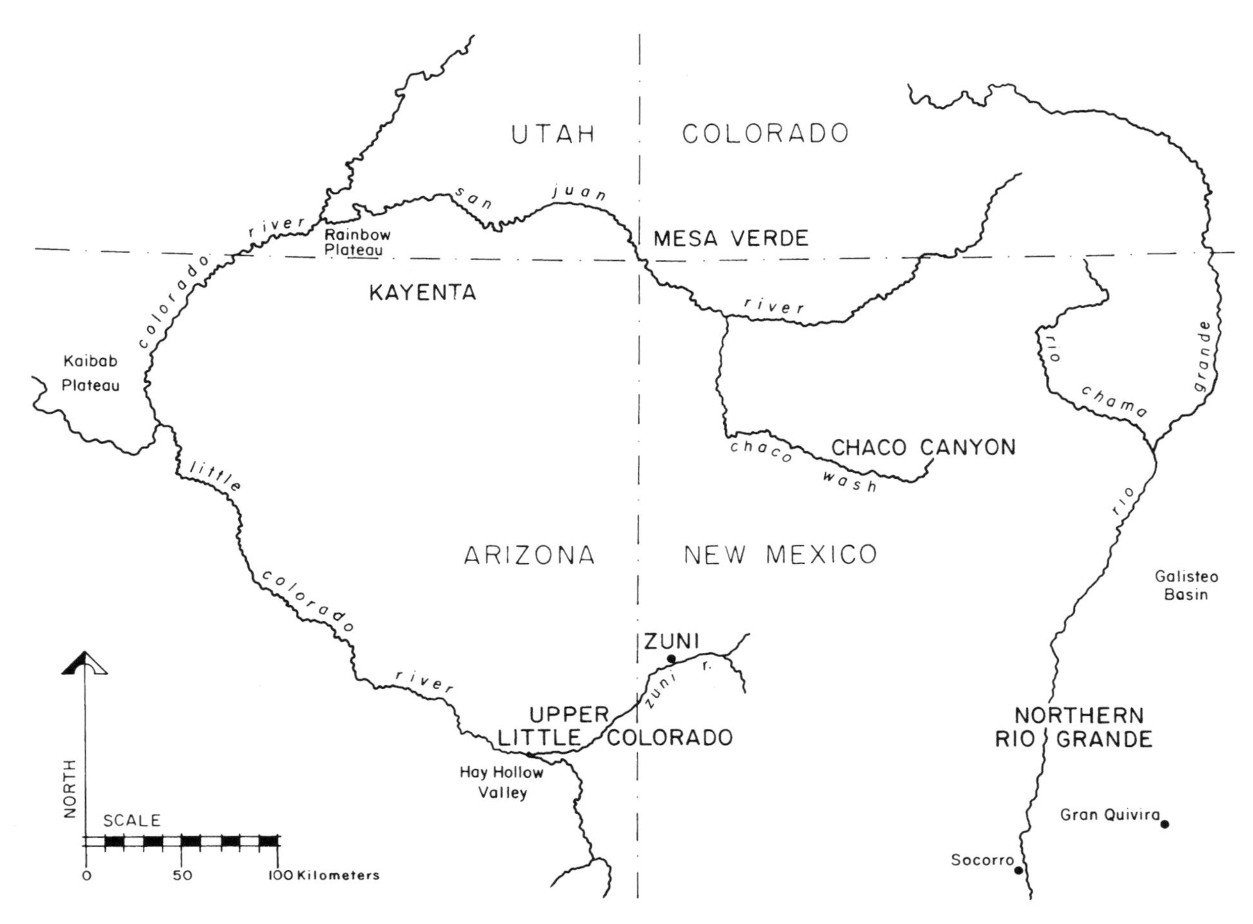


Fig. 9.1 Southwestern localities with reported Anasazi water-control systems.

for Chaco Canyon, New Mexico, a region in which I recently completed a water-control study. I suggest that experimentation with a diversion-type system based on runoff may have been conducive to the development of some form of water-use management. The final portion of this paper is devoted to a discussion of the Chaco material.

Before considering each region in detail, it is appropriate that terminology used be unambiguous. I define *water control* as the myriad of techniques employed in the capture and utilization of water resources. The term *system* is frequently found in discussions of water control in the Southwest and other areas (Evenari et al. 1971). In general, a system is a feature or combination of features that provide water for farming or domestic use. A diversion dam-canal bordered garden unit is an example of a multiple feature system. In some instances one feature constitutes a system, for example, a contour terrace catching runoff or a reservoir collecting water from a spring.

Discussion of water control systems involves reference to water control *features*. In the literature on the Anasazi area, features have not been classified with any degree of regularity. Faced with the choice of continuing to report similar features from diverse areas by different names or attempting some form of standardization, I have chosen the latter course. Both Woodbury (1961*b*) and Hayes (1964) have recently suggested standardized terms. I have tended to follow Hayes where applicable inasmuch as his terms reflect present-day soil conservation terminology. The following feature terms will be used in this paper.

*Bordered Garden:* This feature has been reported as "terrace plots," "grid gardens," "garden plots," "stone-outlined gardens," and "grid borders." All refer to relatively small garden areas enclosed by low earth or stone borders. Bordering served to contain and conserve moisture. Size and shape of individual gardens vary, ranging from single isolated plots to numerous contiguous grids.

*Gravel-mulched Bordered Gardens:* Bordered gardens containing a surface mulch of small gravel added as a moisture retainer.
*Check Dam:* This type of feature has been reported as "trinchera fields," "trinchera plots," and "terraces." Distinction between check dams and contour terraces is occasionally difficult. In general, check dams consist of low rough stone walls built across small intermittent drainages, which frequently occur in ravines or narrow valleys. Usually several check dams are built in one location forming a number of step-like areas; these dams hold both water and soil.
*Contour Terrace:* Contour terraces have been referred to as "linear borders," "boulder bench terraces," and "terraces"; they consist of long rows of low stone walls, or more frequently lines of large stones. Unlike check dams, contour terraces are built across hillsides, talus slopes, or concentrically around small knolls, and are designed to capture slope wash rather than the intermittent drainage of ravines and stream beds. Woodbury (1961*b*:12) uses the term "linear border" for this feature, arguing that the term "terrace" is a misnomer, for "Instead of a step-like profile, . . .the profile is composed of the natural slope of the hillside with regularly spaced stone ridges superimposed on it, each ridge causing only a slight break in profile." I have followed Hayes (1964), who notes that the term "border" has a different meaning in current irrigation farming.
*Ditch:* A narrow shallow cut in the earth for carrying water. A distinction is drawn between ditches and canals; I arbitrarily classify ditches as small cuts less than one meter in width and depth.
*Canals:* A wide deep cut in the earth more than one meter in width and depth used for carrying water.
*Headgate:* A stone or earthen gate controlling the flow of water into a sluice from a canal or ditch.
*Diversion Dam:* A stone or earthen dam designed to temporarily restrict the flow of water for diversion into a ditch or canal.
*Reservoir:* A feature designed for the collection and storage of water. Also reported as "tanks," these features may be totally man-made or may incorporate natural depressions and rock fissures.

## AREAL REVIEW OF ANASAZI WATER CONTROL

### Kayenta

Lindsay (1970:4) reports that water-control devices were noticeably present in the Kayenta during the period from A.D. 1150 to 1300. A variety of water-control features in the Kayenta have been described by several authors (Adams and Adams 1959; Hall 1942; Kidder and Guernsey 1919; Lindsay 1961, 1970; Lindsay et al. 1969; Sharrock, Dibble and Anderson 1961; Schwartz 1960). These features include check dams, contour terraces, bordered gardens, ditches, and reservoirs. As horticultural systems, these features appear in two combinations: check dams and a ditch-contour terrace-bordered garden system (Fig. 9.2). Reservoirs occur singly as domestic water features and at times in combination with the ditch-contour terrace-bordered garden system.

The Cactus Rock Gardens (Lindsay et al. 1969:184) serve as an example of Kayenta check dams. Two adjacent groups of rock rows were placed across the natural slope of an old stream course that provided soil and moisture conservation. The area covered was approximately one-quarter of a hectare. Other check dams were reported by Adams and Adams (1959) along the San Juan and by Hall (1942) for the Walhalla Glades area of the Kaibab Plateau. Although ditches, contour terraces, and bordered gardens have been described as individual features in the Rainbow Plateau area of the Kayenta, there is sufficient evidence (Lindsay et al. 1969:136; Lindsay 1961:174-87) to postulate that these features occurred in combination. This type of system involved the delivery of water by ditches to contour terraces on hillsides and talus slopes with cross walls between terraces forming bordered garden plots (Fig. 9.2). In both the Beaver Creek and Desha Creek communities, contour terraces were traversed by ditches which carried water to the bordered gardens. Bordered gardens excavated at the Desha Creek community had simple water intakes and outlets that led to adjacent plots. An occasional ditch running parallel to the terraces seems to have functioned in distributing water from transverse ditches. Hall (1942) reports a ditch and terrace system from the Walhalla Glades on the north side of the Grand Canyon, but no ditches were excavated, and photographs of the system give the impression that only check dams were present.

Entire systems were limited in size; Kayenta bordered gardens were small, and ranged from one to seven meters square. The Desha Creek community horticultural area consisted of 16 short ditches with associated bordered gardens covering approximately one hectare. The ditch and contour terrace portion of the Beaver Creek community was larger, covering about five hectares.

Lindsay and others (1969:150) report that likely water sources for the ditch-contour terrace-bordered

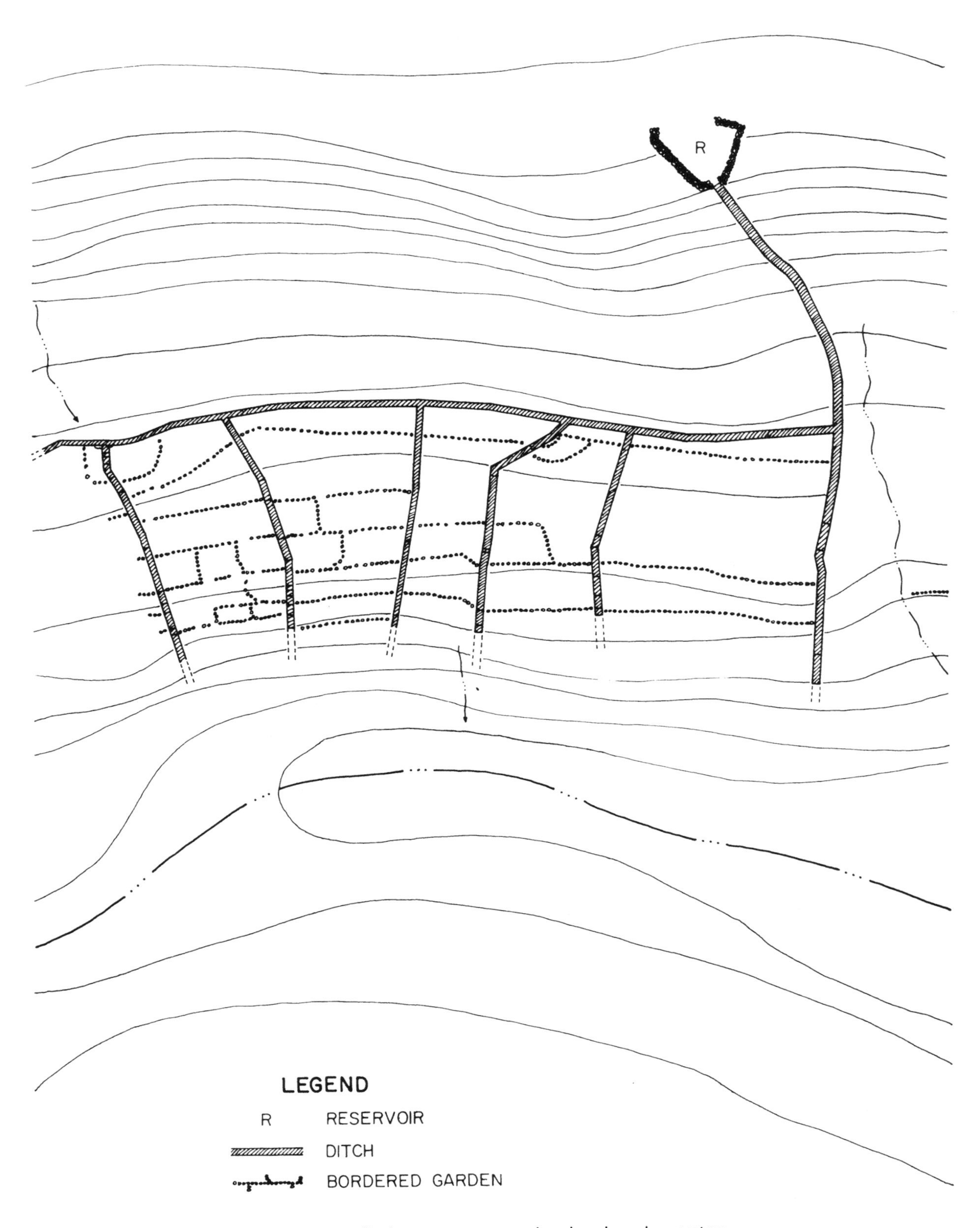


Fig. 9.2 A reconstruction of a Kayenta ditch-contour terrace-bordered garden system.

garden systems in the Rainbow Plateau included "springs, seeps, and surface runoff all emanating from the talus slope above the terrace and ditch systems." Evidence supporting seeps or springs as ditch water sources was found at the Creeping Dune Site in the Rainbow Plateau (Sharrock et al. 1961). A masonry-lined reservoir, similiar to those in other tributary canyons of the San Juan and Colorado rivers, had been built around a spring on a talus slope. An outlet in the reservoir opened into a small ditch that was traced for a short distance down the slope. Attributes of this ditch duplicated those found at Desha Creek and Beaver Creek. The size of ditches found in these locations implies a limited water source, as might be expected from springs or seeps. Slab-lined ditches at Desha Creek averaged 36 centimeters in width and 15 centimeters in depth. Their maximum length was 18 meters, although a ditch at Beaver Creek was traced for about 30 meters. Even though one ditch (No. 15) at the Desha Creek community may have carried water to a number of transverse ditches, it was no larger than any of its outlet ditches. If collection ditches channeled surface runoff, the quantity of runoff was limited. These systems probably were operated in most cases in a manner similar to the historic Hopi Tallahogan Canyon gardens on Antelope Mesa (Hack 1942:34-37). These Hopi garden plots were watered by small ditches drawing water from a spring and a seepage-fed reservoir.

Domestic reservoirs of varied size have been reported from a number of Kayenta areas. These most often consisted of masonry and earthen dams built across water courses to impound runoff, or masonry walls constructed on sloping bedrock to hold surface flow.

## Mesa Verde

Rohn (1970:1) reports that the development of water systems in the Mesa Verde area began possibly as early as A.D. 900, and such devices persisted in use until abandonment at about A.D. 1300. Water-control features present in the Mesa Verde and surrounding area include contour terraces, check dams, ditches, and reservoirs. As systems, contour terraces and check dams occur independently, and in some instances in the same localized watershed. There is some evidence that ditches supplied water to contour terraces and check dams. There is more conclusive evidence, however, that ditches were used for transporting water to reservoirs for domestic use. Domestic reservoirs that trapped runoff from bedrock slopes and ridges, and spring flow from canyon heads have been recorded.

On the basis of present data, check dams appear more common than contour terraces, but as Rohn (1963:445) has commented, "It is quite likely that the majority of such hillside terraces have been partially destroyed or obscured by slope movement." Surveys on Chapin and Wetherill Mesas within the Mesa Verde have indicated that although none of the larger drainages contained check dams, most of the shallow ravines had a series of from three to sixty of these features. Arable land created by check dams was minimal, and Rohn (1963:446) has estimated that the 900 known for Chapin Mesa probably provided only eight to twelve hectares of "top-quality farmland." The Wetherill Mesa survey revealed 136 "terrace" systems, totaling 943 individual "terraces" (Hayes 1964:76). These included contour terraces and check dams. Presumably agricultural land on Wetherill Mesa was no greater than on Chapin Mesa. Although the extent of contour terraces has not been estimated for the Mesa Verde, they were probably as common as check dams. Hayes (1964:79) reports one contour terrace system of approximately one hectare. This appears the average size for a system employing this water control technique.

Runoff provided the water for both check dams and contour terraces. Even though small springs are found at the heads of several of the short canyons and most canyons have some seepage, there is no indication that simple water-diversion techniques were employed for transporting this water to gardens.

Herold (1961:108) has summarized information on five ditches in the Mesa Verde area. The only one reported in detail (Rohn 1963) is the Mummy Lake-Far View Ditch system on Chapin Mesa. The entire system was greater than eight kilometers long. Descriptions of it often make a distinction, however, between the feeder ditch portion north of Mummy Lake and the Far View Ditch to the south of the lake. The ditch began near the top of Navajo Hill and drained ten hectares by means of at least six and possibly nine "catchment feeder ditches." These small ditches were approximately two meters in width and thirty centimeters deep. They led into a gathering basin where the water was diverted into one main feeder ditch. Dimensions of this ditch were similar to the catchment ditches. Water was channeled to Mummy Lake by the feeder ditch and then into the Far View Ditch.

Collection of domestic water is the frequently cited explanation for this entire catchment and diversion system. Mummy Lake served as a domestic source for the Far View ruin group, and Rohn (1963:453) has postulated that the Far View Ditch channeled water "from the wetter northern end of Chapin Mesa to its broader, drier, middle regions primarily for domestic use." Though this function was no doubt important to the system, there is some evidence for distribution to mesa-top fields. For example, a "distributory ditch" tapping the main feeder before it reached Mummy Lake served to transport water to a group of check dams. Both Rohn and Herold also have noted that the Far View Ditch is marked along its course by a number of outlets for field flooding and by several runoff "pick-ups." The Far View Ditch increased notably in size below Mummy Lake (average of nine meters in width and one meter in depth), and should be considered a canal in this portion of the system. This increase in size suggests that additional runoff may have been collected in the vicinity of Mummy Lake. It seems unlikely that water in this ditch was intended for check dams and contour terraces. These systems would have received their own runoff from the mesa top. Instead, the ditch probably serviced mesa-top fields and additional reservoirs.

Data on other ditches in the area are scanty. The Cannonball Ruin ditch carried water to a reservoir; however, reservoirs were not associated with the ditches reported near Yellow Jacket and in Hovenweep Canyon. A possible ditch was distinguished near Lowry Pueblo (Herold 1961:108). Runoff appears as the water source in all these ditches.

Herold (1961:107) has summarized the major characteristics of 15 domestic water reservoirs in the Mesa Verde area. The survey of Wetherill Mesa resulted in the recording of two additional tanks on ridges between Wetherill and Chapin Mesas. Excavated earthen dams and masonry enclosures were built. Only two, Mummy Lake and Cannonball Ruin reservoir, had feeder ditches. The others trapped water from runoff in a manner similar to modern southwestern stock tank collection. These reservoirs varied considerably in size but all were of generous capacity. Mummy Lake was circular, averaging twenty-seven meters in diameter with an original depth of four meters below the high bank. A U-shaped tank on a ridge between Chapin and Wetherill mesas measured eight by twelve meters.

## Northern Río Grande

Water-control structures have been described for the Río Grande Valley and several of its tributary drainages north of Socorro. With a few exceptions (Ellis 1970; Toulouse 1945; Howard 1959), consideration of water control has been casual and descriptive notations are brief (Bandalier 1892; Jeancon 1923; Nelson 1914; Wendorf 1953; and Skinner 1965). Few of the systems can be dated before A.D. 1400, and many date in the early historic period. Granting that most Río Grande water-control systems post-date other Anasazi systems, they are noted here for comparative purposes and because they may reflect earlier systems in the Río Grande.

Features reported for the Río Grande include check dams, contour terraces, gravel-mulched bordered gardens, ditches, and reservoirs. As systems, check dams and contour terraces occur with and without ditches. Bordered gardens, usually with a gravel mulch, occur most often without other features, although Ellis (1970:5) refers to a ditch and bordered garden system at Sapawe. Ditches carrying water to either non-bordered fields or reservoirs are reported, as well as reservoirs without feeder ditches.

Contour terraces and check dams have been noted for late prehistoric or early historic sites at Gran Quivira (Toulouse 1945) and at the sites of Leaf Water and Sapawe in the Chama Valley. To the east at Nambe and Pojoaque, Ellis (1970:4-5) reports informants' references to "terraces" that may be either check dams or contour terraces. Although size is not indicated, the impression is that more land was farmed in other areas. Check dams at Sapawe, according to Ellis, were watered by ditches from El Rito Creek. No such ditch system is reported for Leaf Water, and location of check dams at this site suggests that only runoff would have been available. Check dams at Gran Quivira were watered by ditches draining surplus water out of reservoirs and "artificially built drainage basins." These basins were not dammed as reservoirs, but were instead large areas cleared to bedrock in order to divert slope runoff into terraced fields. Toulouse's map (1945:366) shows that short ditches occasionally were used to transport water from the basins to check dams and other fields.

Gravel-mulched bordered gardens are common in the Río Grande area; they have been noted for the Chama Valley in general (Jeancon 1923; Wendorf 1953) and specifically described at the sites of Sapawe (Ellis 1970:3) and Leaf Water (Wendorf 1953). Ellis further mentions this type of garden on the mesas above the pueblos of Sia and Santa Ana to the south of the Chama Valley. The water source for this type of garden is not firmly established. Both Ellis and Skinner (1965) imply that water may have been diverted in ditches to some plots at Sapawe, but most evidence points to rainfall as the primary source. This system functioned by inhibiting evaporation and preserving moisture through mulching. Ellis (1970:3) describes the Sapawe fields as "edged with stones and mulched with gravel to hold what water fell upon them." Luebben (Wendorf 1953:13) writes of rock enclosures at Leaf Water with interior surfaces covered "with small terrace pebbles of homogeneous size." The size of these systems ranged from small single plots to groups of plots covering several hectares; Skinner (1965:20) describes two plots near Sapawe measuring 15 by 30 meters and 21 by 9 meters. In all cases, rock borders were never more than two stones high.

Ditches at Sapawe (Bandalier 1892; Ellis 1970:4-5) drew water from El Rito Creek and transported it up to four kilometers. Cuts across the ditches revealed a U-shaped profile with an average width of two meters and a depth of one-half meter. Based on informant data, Ellis notes early ditches at Nambe, Pojoaque, Tesuque, San Ildefonso, Santa Clara, and San Juan pueblos; water sources included the Pojoaque, Tesuque, and Río Grande rivers as well as several large springs. Seven ditches are shown on Toulouse's (1945:366) map of Gran Quivira. The channels ranged in length from 21 meters to 320 meters, with shorter ditches being the more common; a trench across one ditch (Howard 1959:88) revealed a width of 5 meters and a depth of 35 to 50 centimeters. The water source was runoff collected from gathering basins and bedrock slopes.

Five circular reservoirs averaging 30 meters in diameter were reported at Gran Quivira; other reservoirs were reported by Nelson (1914:46, 77, 88) in the Galisteo Basin at the sites of San Cristobal, Pueblo Colorado, and Pueblo Blanco. The latter were large semicircular, stone-lined earthern enclosures with surface measurements as great as 275 by 183 meters. In all cases, the water source was runoff.

## Upper Little Colorado-Zuni

Data for this area are limited. Woodbury (1970) has summarized what is known of water control systems in the Zuni area and has remarked that, with few exceptions, descriptions of Zuni systems refer to the historic period. He advises that in addition to assessing the area physiographically, the nature of prehistoric systems may be inferred from ethnohistoric and ethnographic accounts. Plog (1970) has provided some information on systems in three areas of the Upper Little Colorado Valley, dating at A.D. 1000 to 1500. Features defined include ditches, check dams, contour terraces, diversion dams, and bordered gardens but no reservoirs were noted. Contour terraces and check dams appear to have functioned as independent systems; ditches are reported in conjunction with bordered gardens and contour terraces.

Information on system size is scant. Woodbury (1970:2) reports a reference to 162 irrigated hectares near Nutria. Stevenson (1904:351) writes of a much more extensive irrigated area at Ojo Caliente involving 20 to 23 square kilometers, a figure that may be considerably exaggerated. A map of one system in the Hay Hollow Valley of the Upper Little Colorado (Plog 1970:Figure III, 8) shows approximately 20 hectares that may have been irrigated. Ditch dimensions other than length have not been reported. Zuni ditches extended up to 3.2 kilometers, and Plog reports a combined length of about 8 kilometers for the Richville ditch.

Water sources for the Zuni area systems include runoff and large springs. Springs supplying ditch water were reported at Ojo Caliente, Nutria, and Pescado. Plog (1970:7) notes that hillside runoff may have supplied some water to one of the Hay Hollow ditches, but the major water source for the two ditches in this valley was the Hay Hollow Wash. Plog does not indicate if water was permanent or runoff. The source for the Richville ditch is not reported.

## Summary

A variety of features characterize Anasazi water-control systems. Included are ditches and occasionally canals, check dams, contour terraces, bordered gardens, gravel-mulched bordered gardens, and reservoirs. With the exception of ditches and canals, these features occur independently or in several combinations. Ditches and canals always occur with one or more of the other features.

Systems present in the various Anasazi regions are shown in Table 9.1. Kayenta systems include check dams, a combination of ditches, contour terraces, bordered gardens, and possibly runoff-watered contour terraces. Variations on these systems are reported from the Upper Little Colorado-Zuni area and from Mesa Verde. In the Mesa Verde ditch-watered, mesa-top fields probably constituted a variation on the ditch-bordered garden system. A variety of systems are found in the northern Río Grande area (Table 9.1), but the most common one seems to have been the gravel-mulched bordered garden dependent on rainfall. Reservoirs are reported from the Kayenta, Mesa Verde, and northern Río Grande areas.

Three water sources were utilized; rainfall, runoff resulting from rainfall and permanent streams, and springs. Although rainfall contributed to crop watering in all Anasazi areas, the northern Río Grande was the only area where there is evidence for a system designed to make maximum use of direct rainfall. In general, most contour terraces and check dams reported were dependent on runoff. Runoff also provided some water for ditch-field systems at Gran

Quivira, in the Mesa Verde, and possibly in the Upper Little Colorado-Zuni area. Permanent water was utilized less frequently. Water from springs and seeps was collected in small reservoirs for domestic use in some areas and provided crop water through ditching in the Kayenta, northern Río Grande, and Zuni areas. Ditches tapping permanent major streams were reported only from the northern Río Grande. Dating of these Río Grande systems is difficult, and many of the early ditches or *acequias* reported by informants at Río Grande pueblos may be post-Spanish. If these potentially historic ditches are excluded from consideration, the permanent Río Grande sources include only minor streams and springs.

## SYSTEM CLASSIFICATION

Two basic types of systems are apparent: conservation systems and diversion systems. Distinction between the two is based on the method of water use. In conservation systems water is used in place. Examples are check dams and contour terraces watered by runoff, gravel-mulched bordered gardens dependent on rainfall, and reservoirs without feeder ditches. With the exception of spring-fed and seepage reservoirs, Anasazi conservation systems are dependent on rainfall and runoff resulting from rainfall.

Although diversion systems produce water conservation, water is used after it has been collected in one

**TABLE 9.1 Correlation of Water Source, Systems of Control, and System Classification in Five Anasazi Areas.**

| WATER SOURCE and CONTROL SYSTEM | ANASAZI AREAS | | | | | SYSTEM CLASS |
|---|---|---|---|---|---|---|
| | **Kayenta** | **Mesa Verde** | **N. Rio Grande** | **Zuni-Little Colo.** | **Chaco** | |
| **Rainfall** | | | | | | |
| Bordered Garden | X | | X | | X | C |
| Gravel–Mulched B.G. | | | X | | | C |
| **Runoff** | | | | | | |
| Reservoir | X | X | X | | X | C |
| Check Dam | X | X | X | X | | C |
| Contour Ter. | ? | X | X | X | X | C |
| Bordered Garden | | | | ? | | C |
| Ditch–Res. | | X | X | | X | D |
| Ditch–C.D. | | X | X | ? | | D |
| Ditch–C.T. | X | ? | X | X | | D |
| Ditch–B.G. | X | | | | X | D |
| Ditch–Field | | ? | | ? | | D |
| **Permanent Water** | | | | | | |
| Reservoir | X | ? | | | | C |
| Ditch–Res. | | | | | | D |
| Ditch–C.D. | | | X | | | D |
| Ditch–C.T. | X | | X | | | D |
| Ditch–B.G. | X | | ? | | | D |
| Ditch–Gravel–Mulched B.G. | | | X | | | D |
| Ditch–Fields | | | X | X | | D |

B.G. – Bordered Garden
C.D. – Check Dam
C.T. – Contour Terrace

X – Present
? – Questionable presence

C – Conservation System
D – Diversion System

area and transported via ditch or canal to another location. Ultimate use varied and included filling reservoirs as well as the watering of fields and gardens. Unlike conservation systems, Anasazi diversion systems frequently utilized permanent water sources. However, these sources usually produced water in relatively limited amounts. Runoff also was collected in diversion systems, and quantity was dependent largely on the size of the gathering area. In most instances these areas were small. In a few cases, though, notably Chaco Canyon, collection areas were extensive and diversion of runoff became a major task.

## SOME SOCIAL IMPLICATIONS OF ANASAZI CONSERVATION AND DIVERSION SYSTEMS

The review has made it evident that within the Anasazi area runoff was a major water resource (Table 9.1). Although the Anasazi farmer had always utilized runoff, it assumed importance of different proportions sometime after A.D. 900. The shift at this time to a summer-dominant storm pattern made the Anasazi more reliant on summer moisture. As noted earlier, this moisture came predominantly in the form of high energy rainstorms that produced a greater unit runoff rate. Runoff will reflect the rainfall pattern, and thus horticultural systems dependent on runoff will be subject to rainfall variability. Temporal and spatial variability in rainfall can create scheduling problems for the runoff farmer, while variability in the quantity of rainfall can create problems of control and distribution. The prehistoric horticulturists in northern Arizona and New Mexico were faced, therefore, with adjusting to a set of agricultural problems directly related to use of runoff.

Several options were available to the Anasazi for dealing with problems of scheduling, control, and distribution. One method was to employ several kinds of water control techniques, thereby reducing overreliance on the efficiency of a single technique. This option was selected in the Mesa Verde area where check dams, contour terraces, and possibly ditch-watered fields and gardens were all used. Hopi systems reported by Hack (1942) represent a more recent example of the selection of this same strategy.

A second option was to increase the numbers of one type of system without increasing the size of each system or features within the system. Although this method placed reliance on one type of system, the failure of any one unit would not jeopardize the system as a whole. This option was especially important when runoff was utilized, because runoff was divided into manageable lots. In essence, this option prevented overdependence on one large integrated system. Water collection at Gran Quivira exemplified this approach. Water was gathered from approximately nine hectares, but was collected in six artificial drainage basins and then was distributed to fields or reservoirs. The many hectares of gravel-mulched bordered gardens in the Chama Valley serve as another example of this option.

A third option entailed dependence on one type of system, but differed from the above choice in that the size of individual systems was enlarged, and measures for system integration and efficiency were increased. In some instances, selection of this option could have created the need for some form of water management.

Frequently, implications of centralized authority are raised when the question of water management is considered. Millon's (1962) criticism of equating centralized authority with small-scale irrigation is not entirely unwarranted, and many of his remarks are applicable to considerations of southwestern water-control systems. Certainly in most Anasazi systems it was seldom that water source and the system selected for use dictated for management of water resources. For example, centralized authority for water use would have been impractical in most conservation systems. Management of bordered gardens dependent solely on direct rainfall was unnecessary, and conservation systems based on runoff were seldom more than a few hectares in extent, requiring a minimum of maintenance. Just as important, the need for distributing and proportioning water was almost negligible. In general, wetting of most garden areas was assured in check-dam and contour-terrace systems, reducing potential conflict over water rights. Actually, in these systems rights to land could assume more importance. Check dams uppermost in a ravine and contour terraces higher on a hillside would tend to receive a greater percentage of runoff initially; offsetting this possible advantage was the potential hazard of washouts in upper gardens.

Proportioning and distributing water was of some consequence, however, in diversion systems. The diversion of water from *permanent major* streams or rivers can create managerial problems. The construction and maintenance of intake areas, canals, headgates, settling ponds, and other features characterizing large-scale irrigation systems may require varying degrees of organization of a work force. Furthermore, centralized direction may be necessary for efficient and equal distribution of water in such systems.

Diversion systems dependent on permanent water sources are found in a number of the Anasazi regions.

However, with the exception of ditches tapping water from the Río Grande River, systems of the size referred to above have not been reported from the Anasazi Southwest. There is no solid evidence that the Río Grande River ditch systems were functioning prehistorically, and on these grounds they are not considered here. Excluding the historic Río Grande ditches, we are left with a series of systems drawing water from springs, seepage reservoirs, and small streams such as El Rito Creek. The quantity of water was not great, and ditches were relatively small. Taking these factors into account, it is difficult to argue that centralized direction and management of water resources would have been necessary in these systems.

Diversion systems based on runoff have been located at Gran Quivira, Mesa Verde, and Chaco Canyon, and similar systems may have existed in the Upper Little Colorado and Kayenta areas. Only the Chaco data suggest that management of water resources may have become necessary. It appears that the Chacoans, faced with problems of scheduling, control, and distribution as a result of increased runoff, opted for a strategy that involved greater individual system size as well as an increase in efficiency.

Chaco Canyon is located in northwestern New Mexico near the center of the Anasazi area (Fig. 9.1). Water-control studies in the Chaco (Vivian and Mathews 1965; Vivian 1970) have determined that the major system, for which evidence is present, was a runoff-diversion type utilizing diversion dams, canals, ditches, headgates, and earth-edged bordered gardens. Reservoirs tapping water from the canals were probably an additional feature of this system. Evidence for other types of water control in the Chaco is limited to several hectares of contour terracing. Check dams are present in some side canyons and outside the main canyon, but they cannot be distinguished from similar historic Navajo features. All Chaco systems were dependent on runoff. Springs are located in the canyon, but no water-control features are associated with them.

Data gathered from survey and excavation have provided sufficient information for reconstructing the runoff-diversion system. It is known best on the north side of the east-west trending canyon, and because of the topography it may have been extremely limited on the south side. The system was dependent on the gathering of runoff from an inter-cliff zone above the canyon – a zone composed largely of sandstone bedrock sloping gently toward the rim of the canyon. This zone becomes more restricted in area towards the head of the canyon, and presumably because of this factor and other topographic peculiarities, the runoff-diversion system was restricted to the lower 14.5 kilometers of the canyon (Fig. 9.3).

This same portion of Chaco Canyon has been described by Vivian and Mathews (1965:30) as "the most densely populated strip of its size in the Anasazi world." By A.D. 1050 nine large towns of several hundred rooms each, four smaller pueblos ranging from 30 to 100 rooms, and no less than 50 small villages of 10 to 20 rooms (all representing a population of close to 10,000 persons) were crowded into an area 14.5 kilometers in length and averaging one kilometer in width.

Utilizing aerial photographs of the north side of the lower canyon, twenty-eight drainage areas were identified in the inter-cliff zone including three major side canyons that contributed water from an area beyond the inter-cliff zone. The cliff drainages range in size from about 6.5 hectares to 87 hectares, and an average of eight drainages are present between each of the major side canyons. Land drained by the major side canyons is considerably larger. Cly Canyon empties about 526 hectares and Mockingbird Canyon about 810 hectares; the total drainage in the area under consideration comprised approximately 4,250 hectares. Farmland in the lower half of the canyon probably did not exceed 810 hectares; about half of this land was located on the north side and was suitable for runoff diversion farming. Diversion systems have been found in 15 of the drainage areas, including two of the major side canyons.

There is marked similarity in planning and features in these systems. Water was collected from reentrants draining the inter-cliff zone and was channeled into canals by diversion dams. The Rincon-4 North system (Fig. 9.4) of the Peñasco Blanco group serves as a typical example. In this case water collected from the reentrant was directed into a canal (in some parts masonry-lined) averaging 4.5 meters wide and 1.4 meters deep. The canal extended 230 meters from the mouth of the reentrant to a multiple headgate receiving water from the canal and another short ditch collecting local cliff runoff. The gate served to slow water and channel it to a bordered garden area. Information on bordered gardens is derived from aerial photographs of a set of bordered and gridded gardens near the site of Chetro Ketl. Approximately 4.8 hectares of bordered gardens are visible in the photographs, but twice as much land may have been cultivated originally. The 4.8 hectare area is divided into two rectangular plots bordered and separated by canals. Each plot contains 84 bordered gardens, each of which averages 322 square meters in size (23 meters by 14 meters). There are roughly 5.26 gardens to a hectare. Flooding of each garden probably was through temporary openings in the earth borders. The size of plots would have varied depending on available land, size of the drainage area, and the drainage pattern, but it is calculated that as many as 10,000 individual bordered gardens could have been farmed on the north side of the canyon.

The only water for horticulture in the Chaco came from rainfall and runoff. The recent rainfall pattern has been described by Vivian and Mathews (1965:8).

> The Chaco seems to present a pattern in which there are a series of from two to four or five years which are slightly below the mean, separated by single years which are far above the mean. The significance of this, as far as primitive agriculture in the area might be concerned, is that there would be a period of several years in a row in which only 5 inches to 7 plus inches of precipitation were available, followed by single years of destructive rainfall.

With the exception of one early work (Fisher 1934), runoff studies have not been carried out in Chaco Canyon. I have made some preliminary estimates of runoff in the canyon based on the nomogram of rainfall-runoff relation prepared by Evenari (Evenari, Shanan, and Tadmor 1971:146-47) for areas he studied in the Negev. I do not wish to imply close similarity between the Chaco and the Negev; however, I would like to make it clear that runoff studies could provide invaluable information for southwestern archaeologists. The Chaco estimates are speculative, but they do suggest that runoff was of major importance to Chacoans faced with the necessity of adjusting to a shift in the precipitation pattern.

For this discussion I have calculated runoff for the previously cited Rincon-4 North system of the Peñasco Blanco group. These estimates are based on the assumption that watersheds were not cleared or specially prepared to increase runoff. Present area available for farming in the Rincon-4 North system comprises 9.7 hectares; a figure that is probably close to the prehistoric conditions. If the Chetro Ketl system is used as a model for garden size, we could expect a farm of four plots with a total of 336 individual bordered gardens. Based on data gathered from all canyon systems, it is probable that minimal features in this one system would have included a diversion dam, primary canal, primary multiple gate, six field ditches, four secondary gates, and 336 bordered gardens.

The drainage area for this farm is approximately 44.5 hectares. Potential runoff was figured on rainfall for the months of May through September for the years 1956, 1967, and 1971. One of the driest years in recent history in Chaco was 1956, when the total annual precipitation was 8.55 centimeters. Summer rainfall for this year was 3.70 centimeters. The summer of 1967 was considerably wetter, with 19.65 centimeters of rain. Somewhat less water was available in 1971, when 14.55 centimeters was recorded. If all of this recorded moisture had produced runoff, the Rincon 4 North farm would have received approximately 44,970 decaliters (118,800 gallons) in 1956, about 1,124,234 decaliters (2,970,000 gallons) in 1967, and 994,356 decaliters (2,494,800 gallons) in 1971. Actually, much of the recorded rainfall did *not* produce runoff.

Although weather records for the canyon do not always provide information sufficient for determining quantity of runoff, they do reveal the extreme variation in quantity of daily rainfall and in period of time that rain fell. Both factors effect runoff. For example, daily rainfall in 1956 was so minimal that runoff would not have occurred, even though total precipitation for the period, if considered as one unit, would have accounted for some runoff. Even significant amounts of daily rainfall are not necessarily indicative of runoff. Though records for 1967 and 1971 show a number of days with precipitation sufficient for producing runoff, this daily record often represents a total of several periods of rainfall throughout 24 hours. Rainfall was not measured for each period, thus making it difficult to determine if runoff may have occurred during any one of the periods. To illustrate, the 4.65 centimeters of rain that fell on September 29, 1971, included at least two separate precipitation periods totaling two hours, and may have included late evening rainfall as well. It should be noted that this rainfall would have occurred too late in the season to have been of much value to crops. Most Chaco summers are characterized, however, by several intense thundershowers that produce great quantities of runoff that quickly drain out of the canyon. This fact is supported by the weather records and personal observation. For instance, the storm of July 22, 1967, produced 3.04 centimeters of rain in a one-hour period. This rain

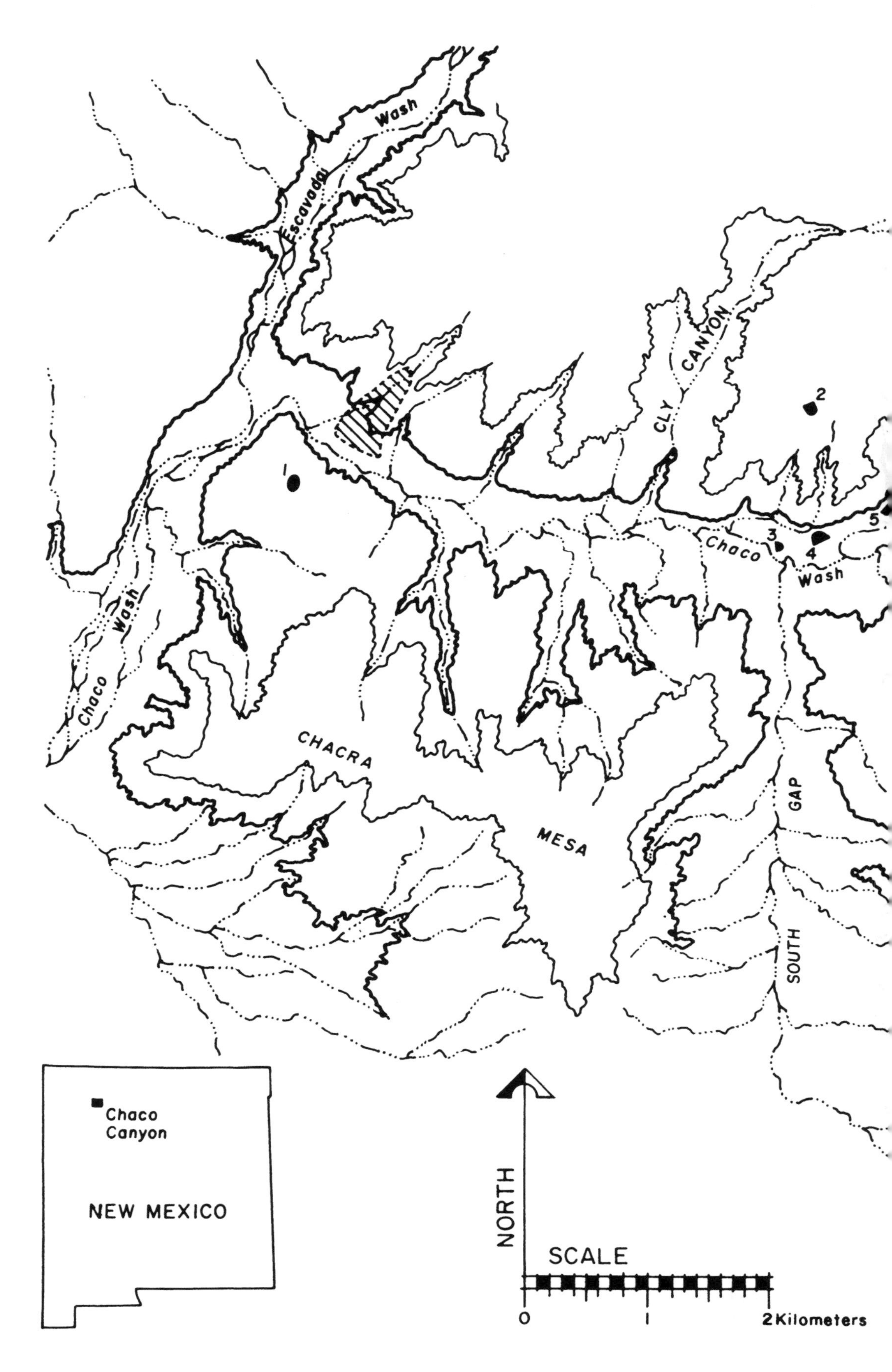
Escavada
Wash
CLY
CANYON
Chaco
Wash
Chaco
Wash
CHACRA
MESA
GAP
SOUTH
1
2
3
4
5
Chaco
Canyon
NEW MEXICO
NORTH
SCALE
0
1
2 Kilometers

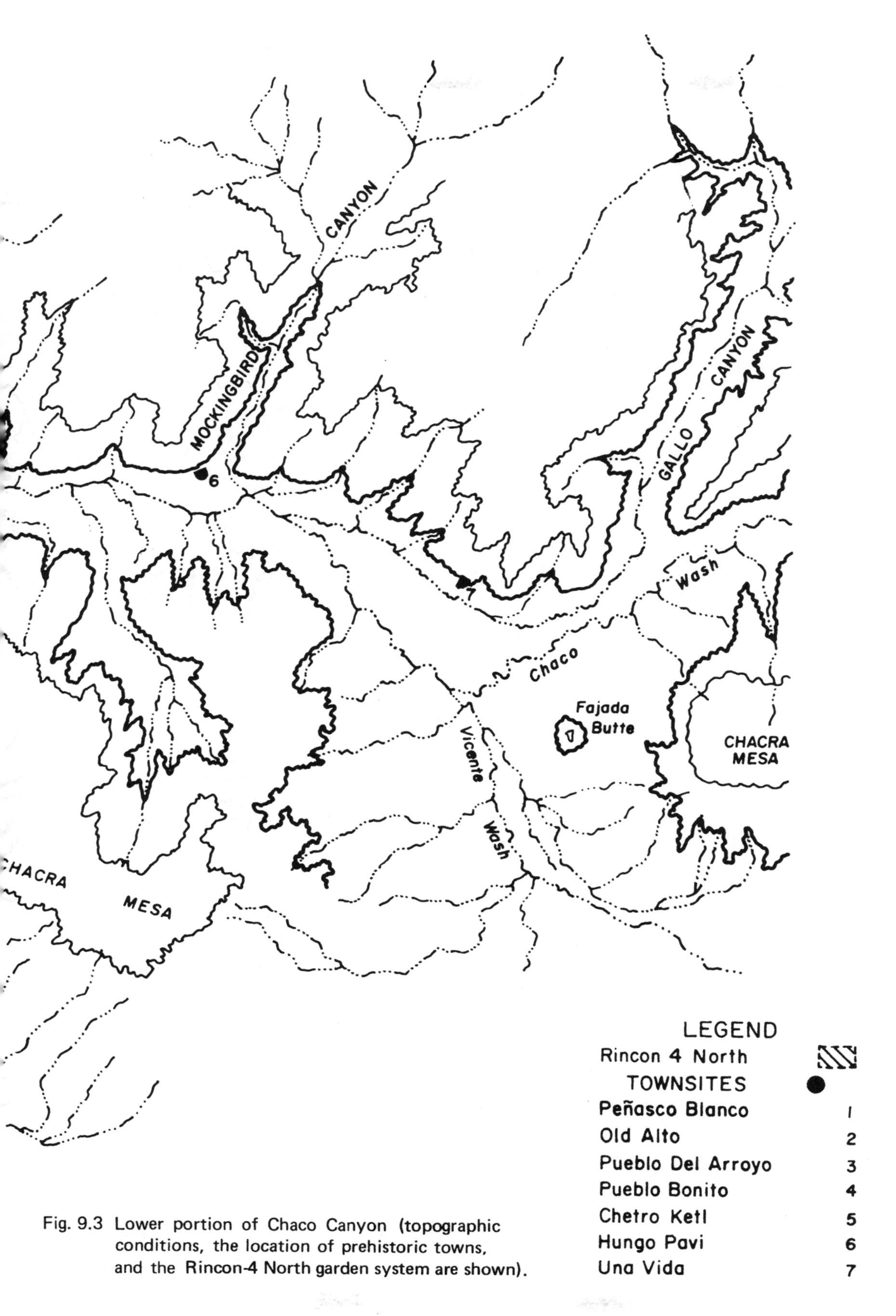


Fig. 9.3 Lower portion of Chaco Canyon (topographic conditions, the location of prehistoric towns, and the Rincon-4 North garden system are shown).

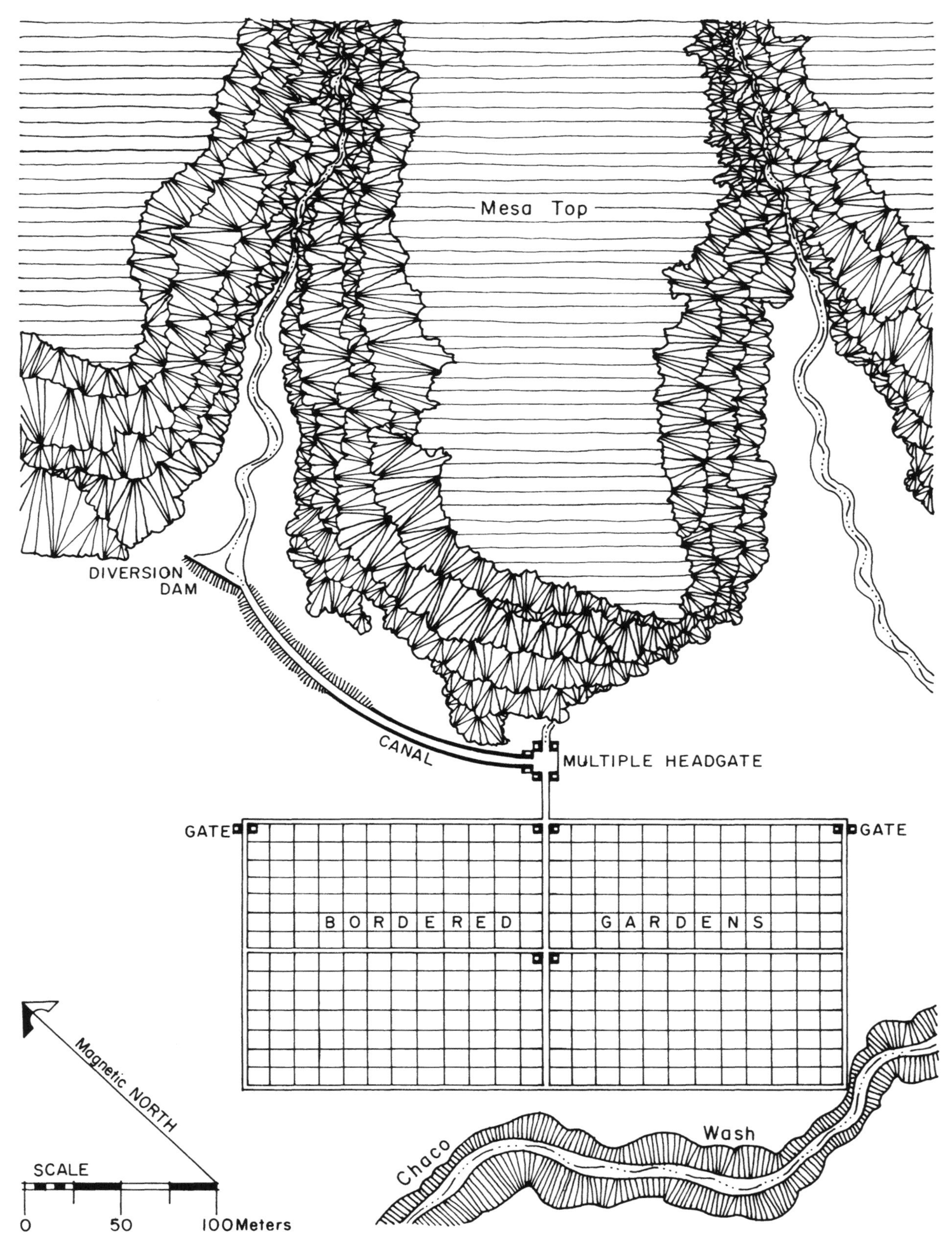


Fig. 9.4 Rincon-4 North water-control system in Chaco Canyon (a typical Chaco system utilizing diversion dam, canal, gate complex, and bordered gardens).

would have yielded approximately 204,000 decaliters (540,000 gallons) of water for the 9.7 hectares of bordered gardens in the Rincon-4 North system had this farm been in use. Presumably proportionate amounts of water were available for most of the other canyon systems as a result of this one storm.

In order to have utilized this moisture effectively, a system for the capture and control of runoff was necessary. The system had to be compatible with high population density, relatively limited acreage for farming, and the unpredictability of runoff in terms of occurrence, quantity, and intensity of flow. It might be argued that the Chaco systems that were developed represent only an elaboration on the second option discussed earlier – to increase the numbers of one type of system without increasing the size of each system or features within the system. I would argue instead that the Chaco Anasazi selected the third option, which entailed dependence on a single type of system and involved an increase in system size and elaboration of measures for system integration and efficiency. There are some supportive data for my argument. The network of diversion systems developed by the Chacoans, for example, was expanded to encompass virtually all land on the north side of the lower canyon, and each system included the maximum area of land available. Methods for increased efficiency included the gridding of fields for equal proportionment and conservation of water, and standardization in form and function of system features. Despite the fact that the drainage pattern in the canyon precluded the development of a single canal system, the individual drainage areas were integrated into one pattern that made maximum use of all land and water resources.

If system size were enlarged, and measures for efficiency and integration increased, was there a need for water use management in the Chaco? I have proposed elsewhere (Vivian 1970) that farmland on the north side of the canyon was held as corporate property by the various towns in the lower canyon. Presumably a number of drainage areas were held by each of the towns. With a shift in precipitation pattern, the development of systems for conservation of water would have assumed increasing importance for the leaders and members of each community. Ironically, the substance that made the land valuable – water – was also a major threat to the land. One consequence of the summer-dominant storm pattern was the headward erosion of arroyo cutting, producing a lowering of water tables and loss of farm land. A critical concern of the Chacoans, then, was loss of land through erosion so that control of floodwaters provided not only for *water conservation* but *land preservation* as well. I would postulate that through time prime farm land in the canyon became a commodity of great value, and that once it was acquired and developed, its protection became a community responsibility. High land value, limited area, irregular moisture pattern, and the continual threat of land loss could have contributed to the evolution of some form of centralized authority to protect communal property. The means by which management was carried out is not necessarily critical to this discussion. Rather, I would stress that the potential for centralized management of water resources was present in Chaco Canyon, and would suggest that this potential may be greater in systems based on diversion of runoff than in other types of small-scale irrigation.

## REFERENCES

ADAMS, W. Y. and N. K. Adams

1959 An Inventory of Prehistoric Sites on the Lower San Juan River, Utah. *Museum of Northern Arizona Bulletin* 31. Flagstaff.

BANDALIER, Adolf F.

1892 Final Report of Investigations among the Indians of the Southwestern U.S. Carried Out Mainly in the Years from 1880 to 1885. Part II. *Papers of the Archaeological Institute of America.* Cambridge.

ELLIS, Florence H.

1970 *Irrigation and Water Works in the Rio Grande.* MS. 1970 Pecos Conference Water Control Symposium.

EVENARI, Michael, Leslie Shanan and Naphtali Tadmor

1971 *The Negev: The Challenge of a Desert.* Cambridge: Harvard Univ. Press.

FISHER, Reginald G.
1934 Some Geographic Factors that Influenced the Ancient Populations of the Chaco Canyon, New Mexico. *The Univ. of New Mexico Bulletin, Archaeology Series,* Vol. 3, No. 1, Whole Number 244. Albuquerque: Univ. of New Mexico Press.

HACK, John T.
1942 The Changing Physical Environment of the Hopi Indians of Arizona. *Papers of the Peabody Museum, Harvard Univ.* Vol. 35, No. 1.

HALL, E. T. Jr.
1942 Archaeological Survey of Walhalla Glades. *Museum of Northern Arizona Bulletin* 20. Flagstaff.

HARRIS, Arthur H., James Schoenwetter, and A. H. Warren
1967 An Archaeological Survey of the Chuska Valley and the Chaco Plateau, New Mexico. Part I, Natural Science Studies. *Museum of New Mexico Research Records,* Number 4. Santa Fe.

HAURY, Emil W.
1965 Snaketown: 1964-1965. *The Kiva.* 31:1-13. Tucson.

HAYES, Alden C.
1964 The Archeological Survey of Wetherill Mesa, Mesa Verde National Park, Colorado. *National Park Service Research Series* No. 7A. Washington, D.C.

HEROLD, Joyce
1961 Prehistoric Settlement and Physical Environment in the Mesa Verde Area. *Univ. of Utah Anthropological Papers,* No. 53. Salt Lake City.

HOWARD, Richard M.
1959 Comments on the Indian's Water Supply at Gran Quivira National Monument. *El Palacio.* 66:85-91. Santa Fe.

JEANCON, J.A.
1923 Excavations in the Chama Valley, New Mexico. *Bureau of American Ethnology Bulletin,* 81. Washington, D.C.

KIDDER, A. V. and S. J. Guernsey
1919 Archaeological Explorations in Northeastern Arizona. *Bureau of American Ethnology Bulletin* 65. Washington, D.C.

LINDSAY, Alexander J., Jr.
1961 The Beaver Creek Agricultural Community on the San Juan River, Utah. *American Antiquity.* 27:174-87. Salt Lake City.
1970 *A Descriptive Summary of Soil and Water Conservation Devices and Settlement Patterns in the Kayenta Anasazi Region.* MS. 1970 Pecos Conference Water Control Symposium.

LINDSAY, Alexander, J., Jr., J. Richard Ambler, Mary Anne Stein and P. M. Hobler
1969 Survey and Excavations North and East of Navajo Mountain, Utah, 1959-62. *Museum of Northern Arizona Bulletin* No. 45. Flagstaff.

MILLON, Rene
1962 Variations in Social Responses to the Practice of Irrigation Agriculture. In *Civilizations in Desert Lands,* edited by Richard B. Woodbury. *Univ. of Utah Anthropological Papers,* No. 62. Salt Lake City.

NELSON, Nels C.
1914 Pueblo Ruins of the Galisteo Basin, New Mexico. *Anthropological Papers of the American Museum of Natural History,* Vol. 15, Part 1. New York.

PLOG, Fred
1970 *Water Control in the Upper Little Colorado.* MS. 1970 Pecos Conference Water Control Symposium.

ROHN, Arthur H.

1963 Prehistoric Soil and Water Conservation on Chapin Mesa. *American Antiquity,* Vol. 28, No. 4. Salt Lake City.

1970 *Social Implications of Pueblo Water Management in the Northern San Juan.* MS. 1970 Pecos Conference Water Control Symposium.

SCHOENWETTER, James

1970 Archaeological Pollen Studies of the Colorado Plateau. *American Antiquity.* 35:35-48. Salt Lake City.

SCHOENWETTER, James and A. E. Dittert, Jr.

1968 An Ecological Interpretation of Anasazi Settlement Patterns. In *Anthropological Archaeology in the Americas,* ed. B. J. Meggers, pp. 41-66. Anthropological Society of Washington, Washington.

SCHOENWETTER, James and Frank W. Eddy

1964 Alluvial and Palynological Reconstruction, Navajo Reservoir District. *Museum of New Mexico Papers in Anthropology,* No. 13. Santa Fe.

SCHWARTZ, Douglas W.

1960 Archaeological Investigations in the Shinumo Area of Grand Canyon, Arizona, *Plateau.* 32:61-67. Flagstaff.

SHARROCK, Floyd W., David S. Dibble and Keith M. Anderson

1961 The Creeping Dune Irrigation Site in Glen Canyon, Utah. *American Antiquity.* 27:188-202. Salt Lake City.

SKINNER, S. Alan

1965 A Survey of Field Houses at Sapawe, North Central New Mexico. *Southwestern Lore.* 31:18-24. Boulder, Colorado.

STEVENSON, M. C.

1904 The Zuni Indians. *23d Annual Report of the Bureau of American Ethnology.* Washington.

STEWART, G. R.

1940 Conservation in Pueblo Agriculture: I, Primitive Practices; II, Present-day Flood Water Irrigation. *Scientific Monthly,* 51:201-20, 329-40. New York.

STEWART, G. R. and Maurice Donnelly

1943 Soil and Water Economy in the Pueblo Southwest: I, Field Studies at Mesa Verde and Northern Arizona; II, Evaluation of Primitive Methods of Conservation. *Scientific Monthly,* 56:31-44, 134-44. New York.

TOULOUSE, Joseph H., Jr.

1945 Early Water Systems at Gran Quivira National Monument. *American Antiquity,* 10:362-72. Menasha.

VIVIAN, R. Gwinn

1970 An Inquiry into Prehistoric Social Organization in Chaco Canyon, New Mexico. In *Reconstructing Prehistoric Pueblo Societies,* ed. William A. Longacre, pp. 59-83. Albuquerque: Univ. of New Mexico Press.

VIVIAN, R. Gordon and Tom W. Mathews

1965 Kin Kletso, a Pueblo III Community in Chaco Canyon, New Mexico. *Southwestern Monuments Association Technical Series,* Vol. 6, Part 1.

WENDORF, Fred

1953 Salvage Archaeology in the Chama Valley, New Mexico. *School of American Research Monographs,* No. 17. Santa Fe.

WOODBURY, Richard B.

1960 The Hohokam Canals at Pueblo Grande, Arizona. *American Antiquity.* 26:267-70. Salt Lake City.

1961*a* A Reappraisal of Hohokam Irrigation. *American Anthropologist.* 63:550-60. Menasha.

WOODBURY, Richard B.

1961*b* Prehistoric Agriculture at Point of Pines, Arizona. *Society for American Archaeology, Memoirs,* No. 17. Salt Lake City.

1962 Effects of Environmental and Cultural Limitations Upon Hohokam Agriculture, Southern Arizona. In *Civilizations in Desert Lands*, ed. Richard B. Woodbury, pp. 41-55. *Univ. of Utah Anthropological Papers,* Number 62. Salt Lake City.

1970 *Symposium on Prehistoric Southwestern Water Control Systems: The Zuni Area.* MS. 1970 Pecos Conference Water Control Symposium.

CHAPTER 10

# IRRIGATION AND MOISTURE-SENSITIVE PERIODS A Zapotec Case

Theodore E. Downing

*Bureau of Ethnic Research and Department of Anthropology, University of Arizona*

Cultigen growth depends upon appropriate amounts of heat, sunlight, carbon dioxide, soil nutrients, space, time, and moisture. Of these, modification of moisture content has been one of man's more common tools for increasing cultigen yields. Given other adequate inputs, cultigen productivity is optimized when critical amounts of water are applied at the correct time. Herein, I discuss the importance of the amount, source, and timing of water application upon an irrigation system in southern Mexico.[1]

## THE OAXACA VALLEY

The Oaxaca Valley lies at 5000 feet, 250 miles southeast of Mexico City. The Río Atoyac traverses the northwestern and southern arms of the Y-shaped valley. The Río Salado, bisecting the southeastern arm, joins the Atoyac near Oaxaca City at the Y's center. Extremely rugged mountains surround the valley, whose northern rim forms the continental divide.

Lying in the rain shadow of the divide, this watershed is a pawn of the Caribbean trade winds. In the dry season, roughly from October to April, stream beds are filled with sand, rocks, and boulders – but no water. Spring trade winds push rain-laden clouds onto the divide; these early clouds break away, bringing sporadic cloudbursts to a few fields in the valley. High on the escarpment slopes, heavy downpours rejuvenate dormant stream beds. Torrents of brown, foaming waters wash rich vegetable matter into the valley.

In July, daily rains punctuate the afternoons. A short period of decreased precipitation in August is followed by heavy rains in September. In early October, the final drizzles dampen the high slopes, and another dry season begins (Figure 10.1).

Rainy seasons are predictable; their magnitudes and starting and ending times are not. The wide range of annual rainfall, varying from 435 to 825 millimeters, makes the average annual rainfall (673 mm.) a meaningless statistic (Schmeider 1930:7). Valley farmers recognize this variability and contrast "wet" with "dry" years, occasionally comparing these with a mythical "average" year that rarely occurs.

Precipitation's random spatial distribution adds more uncertainty to farming. Most rains occur as cloud bursts or severe local thunderstorms which drench a small area – often no larger than a few square kilometers – while adjacent lands remain dry.

Rainfall's varying annual magnitude and uneven spatial distribution combine to make dry farming a risky enterprise. Moreover, these meterological conditions keep stream flows intermittent and of varying volumes, making irrigation from a temporally stable, permanent surface water source physically impossible.

## VALLEY IRRIGATION

Flannery, Kirkby, Kirkby, and Williams (1967) outline the valley's long history of irrigation, discerning evidence of irrigation in one form or another since the Early Formative (1200 to 900 B.C.). Further, they distinguish three types of small-scale irrigation: *pot, canal,* and *floodwater. Pot irrigation* involves hand mining of groundwater tapped by means of small, shallow wells. Primitive well digging technology limits this technique to those parts of the valley with relatively high water tables (1.5 to 3.0 meters). *Canal irrigation* is gravity-flow diversion of perennial streams or springs, while the diversion of occasional flash floods during the rainy season is called *floodwater irrigation.*

It is important to realize that contemporary Oaxacan irrigation systems may not show historical continuity with pre-Columbian systems. Many of the canal and floodwater irrigation systems located on the high piedmont were abandoned following the Conquest. This abandonment was the result of a

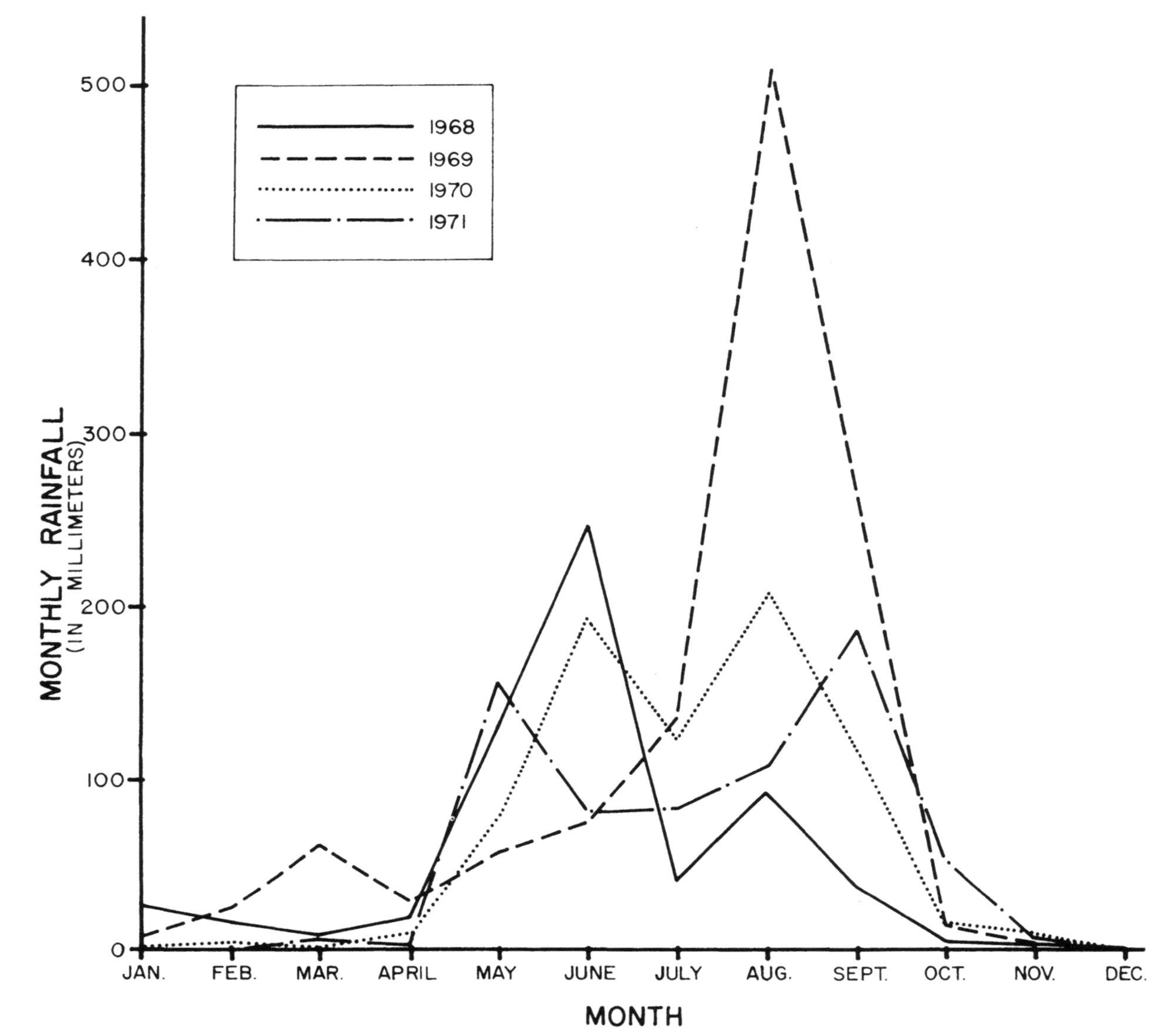


Fig. 10.1 Rainfall in the Tlacolula Valley, 1968-71. Source: gauging station of Cecil Welte, Oficina de Estudios de Humanidad del Valle de Oaxaca, Oaxaca, Mexico.

combination of secondary effects of depopulation, decreased demand for food, and consequent disruption of local level, socio-political organization. The late seventeenth-century population expansion brought resettlement of the high piedmont and re-establishment of some of the irrigation techniques. Presumably, these irrigation systems are similar to preconquest systems, although this assumption demands more careful historical reconstruction than has been previously attempted.

Today, small-scale irrigation systems dot the Oaxaca Valley floor. *Small-scale* implies that the lowest level in the Mexican political hierarchy, the municipality, designs, builds, maintains, and administers the technique. These systems show considerable variation in allocation, administration, and physical size (cf. Lees 1973).

## DÍAZ ORDAZ

The town of Díaz Ordaz, also known as Santo Domingo del Valle, operates one of these small-scale irrigation systems. Located in the Tlacolula wing, this nucleated community of 4,000 Zapotec-Spanish-speaking peasants borders the village of Mitla (Parsons 1936).

Resettled during the sixteenth century, Díaz Ordaz's economy was derived from wage labor on a nearby hacienda and herding, supplemented by dry farming and blanket weaving. This economy was altered by the political and economic disruptions in the early twentieth century. The Revolution forced the community to fend for itself through acceptance and a progressive expansion of a combined canal-floodwater irrigation system. Today, Díaz Ordaz ranks as one of the few corn-exporting villages in the Tlacolula arm's high piedmont. Irrigated agriculture is supplemented by wage labor and, to a lesser extent, sheep and goat herding, blanket weaving, and other small crafts.

The village cultivates less than 10 percent of its estimated 50 square kilometers of land. Most village lands are mountainous and barren, suitable for firewood, maguey, or beans. Farming proves most profitable in the small corner of village land on the valley floor. Of this land, approximately 500 hectares are dry farmed, yielding at best one maize crop per year. Other dry farmed crops include maguey, various kinds of beans, squash, and some castor beans.

Only a fraction of Díaz Ordaz's lands are regularly irrigated (±150 hectares). This land might be called the community's "breadbasket." Not only does it offer two crops of maize per year, but also it permits bountiful yields of alfalfa, winter wheat, chick peas, beans, squash, and castor beans.

## WATER SOURCES

Farmers divert irrigation water primarily from an intermittent stream called the *Heu Ro'o* ("Large Stream"). Its watershed extends to the crest of the continental divide, encompassing lands of the upstream settlement, San Miguel del Valle.

San Miguel's agricultural practice, the swiddening of beans and potatoes, plays a critical supportive role for crops in Díaz Ordaz. Floodwater used by Díaz Ordaz farmers derives its rich nutrient content from vegetable material cleared by San Miguel farmers from their fields and subsequently carried downstream by erosion. Furthermore, San Miguel's peak work season is different from that of Díaz Ordaz, thereby providing the latter with a cheap source of labor for their more intensive agriculture.

Díaz Ordaz's Zapotecs segregate two kinds of river water (*nis heu*). The first, flash floods (*nis bae'n*), are infrequent and limited to one to six annual flows which last from two to six hours. In the daytime, Zapotecs attempt to predict these flows by observing rainfall patterns in the river's watershed. This prediction often provides irrigators with a two-hour advance warning. At night, only a distinctive rumble of flowing, turbulent water forewarns floodwater's arrival.

Day or night, when floodwater arrives, all social activities are secondary to channeling this highly valued water onto fields. For instance, I witnessed the local elementary school director lose his audience during graduation exercises. Floodwater arrived during his lengthy, prepared speech. Impatient fathers rolled up their pant legs, and at the first pause in the address, excused themselves and ran to get shovels. Manning their sluice gates had precedence over this otherwise important occasion.

The second category, clear water (*nis nia*), includes all river flow that is not floodwater. Clear water occurs intermittently throughout the rainy season – originating as either precipitation runoff or groundwater seepage in the river bed.[2]

Zapotecs distinguish four parts of a canal[3] (Figure 10.2). The first of these, *ru'u tom* (literally, "drinking mouth"), might best be glossed "diversion dam." Six diversion dams take water from the Large Stream. These dams stand 2 meters in height and 20+ meters in width; they are constructed of logs spaced every meter, interlaced with saplings, rock, and earth fill. Half the dams rest upon concrete footings, which were recently built to retard the persistent undercutting problem.

During floodwater peaks, these dams channel water down their main canals. Each irrigator along a canal regulates the amount of water entering his fields by means of a breach in the bank plugged by a large rock. As irrigators close their sluice gates, back pressure builds up and combines with the massive force of flooding and undercutting. The diversion structure collapses. Floodwater rushes downstream and is again diverted by the next dam. This process continues until either the flood is exhausted or all fields are irrigated.

In addition to diverting floodwater, the diversion dams impound "clear water" during periods of minimal flow. In this case, the main canal sluice gate is closed until the small reservoir behind the dam reaches its capacity. Then, the gate is opened and one "tank" or *reigo* released.

The *ja'a tom* is that stretch at the canal's head located between the main sluice gate and the first irrigated parcel. Like the diversion dam, it is subject to heavy silting and requires periodic maintenance.

The remaining length of the canal is subdivided into segments called *bias* plus a proper name. These segments have no operational significance in the irrigation system. Rather, they are historical survivals marking the progressive extension of a given canal. Díaz Ordaz's canal network has 18 segments, suggesting at least as many distinct building phases since the turn of this century. Extensions were organized by some individual or group at the end of an existing canal. Using only a small experimental ditch, they would test the feasibility of a canal by gaining permission from all owners whose lands the ditch would pass, and then observe if water moved correctly in the ditch. If it did, they would organize a cooperative effort by all parcel holders who would benefit from an extension. Costs were distributed among farmers in proportion to the size of their land holdings.

Topographically, extensions appear to be limited by natural obstacles, like arroyos. Crossing an arroyo with an irrigation canal requires a considerable outlay in manpower and materials for a small aqueduct (*calicanto*). Disagreements invariably arise between up-canal irrigators as to why they should contribute to the costs of aqueducts which benefit down-canal irrigators. The solution: stop the extension at the arroyo and let the farmers down-canal organize their own group since they would all directly benefit from their investment.[4] Next, through a process which I poorly understand, the segment's (bias) irrigators subdivide themselves into anywhere from 2 to 12 sub-units called *tramos.*

## TRAMOS

Tramos (sections) are the most significant spatial and administrative unit within this irrigation system. A section encompasses all parcels irrigated by a common sluice along the main canal. Each household owning land within this spatial unit is called a *regante* or irrigator. The number of irrigators in a section varies from 13 to 41, with a median of 28. Municipality records of sectional membership suggests that this ratio has remained constant since the 1930s, despite increases in the number of sections in the village.

Section membership is obligatory and contingent upon ownership and/or cultivation of a parcel within the designated territory. Transfer of ownership, either temporarily through sharecropping or permanently through inheritance or sale, necessitates transfer of section membership.

Each section has administrative officials: president, scribe, and treasurer. Succession to these offices comes through rotation, each parcel holder taking an office for one year. However, personnel may overlap: a woman or man may be an official in more than one section at one time. This situation is contingent, of course, upon owning dispersed parcels in village lands.

The section officials have duties explicitly detailed in the group's charter. They mediate disputes among their unit's irrigators, and hopefully, resolve them informally before escalation to higher authorities.

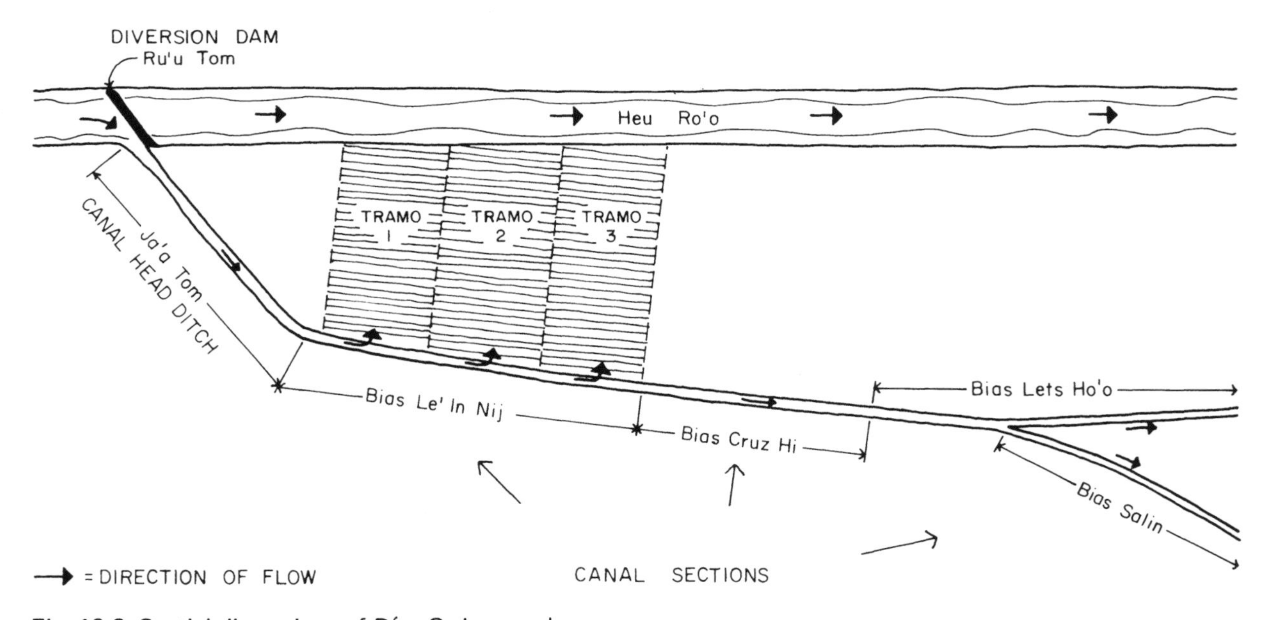


Fig. 10.2 Spatial dimensions of Díaz Ordaz canals.

They assign obligatory work loads to all irrigators within the section, collect a small tax, and distribute water allocated to their section. For these duties the officials receive no compensation and little praise.

Work loads (*tareas*) and taxes (*impuestos*) are distributed among water users proportionate to the amount of maize which potentially could be seeded on their parcels at one planting.[5] Each section along a canal contributes equal shares either in labor, materials, or more rarely, cash, for the upkeep of the canal head (*ja'a tom*), and the diversion dam. Major repairs to aqueducts or to breaks in the main canal are borne by those downstream from the damage. In contrast, work within the section is not done cooperatively. Section officials mark off lengths of the canal to be cleaned and assign them to different irrigators. These irrigators or their hired labor must clean their assigned portions by a certain date, but do not necessarily work together.

There is considerable variation in water allocation among sections. Within some sections the total allotment of water is divided equally by the users. Others apportion water according to the surface area of land held by each irrigator. Sections receive allocations of water from the next level in the irrigation hierarchy, the *síndico.*

## SÍNDICO

The síndico is a local authority who is in charge of water distribution to the 70+ irrigation sections (*tramos*). Water allocation is not the síndico's primary duty. As the second highest official in the municipality's local level political system (known to Meso-American scholars as the civil-religious hierarchy), the síndico is the jural representative of the state government at the local level. He shares responsibility with the local judges in the adjudication of all intra-village disputes, and holds sole responsibility for referring all cases of capital offenses and "civil disobedience" to the district courts. The vague definition of this latter offense provides the síndico with his power. Extreme disagreement of an irrigator with others may be interpreted by the síndico as endangering community harmony. Civil disobedience, which may be interpreted as extreme counter revolutionary activity may lead to heavy legal expenses and possible lengthy incarceration. Villagers are aware of this sanction although the síndico seldom exercises it. This power would lead to a potentially dictatorial situation if it were not for the obligatory and limited term of office. Like other municipal officials, the síndico is chosen by consensus and serves an economically and psychologically punishing three-year term. Like the tramo officials, he receives neither compensation nor praise.

Criteria for selection of síndico does not include water administration abilities or irrigation experience. The past two síndicos in Díaz Ordaz have considered water administration a nuisance and readily admitted they poorly understood the irrigation system beyond their own duties. These duties are to 1) inspect the diversion dams and head sections and decide when they are in need of repair, 2) mediate disputes not resolved by section (tramo) officials, and 3) allocate water among sections. Understanding how water is allocated requires considering temporal changes in the availability of water.

## TIME, WATER AND RIGHTS

Western European law distinguishes three principal doctrines for the distribution of surface water rights. Under riparian doctrine, owners of land located adjacent to a water source have rights to it whenever they wish. This doctrine is modified under the condition called "correlative rights" wherein these prime users should exercise some consideration of others; e.g., downstream users' rights should not be completely usurped by upstream users. Finally, under the prior appropriation doctrine, priorities to water are determined by historical precedence or administrative fiat rather than by proximity of an owner's land to the source. Kazmann (1965:122) and others note that the riparian doctrine corresponds with areas of excess water, like the eastern United States, and the prior appropriation doctrine corresponds with water-deficient regions, like the southwestern United States.

The Mexican state follows the doctrine of prior appropriation. Díaz Ordaz Zapotecs selectively ignore the doctrine and alternate between riparian and prior appropriation rights within the same year.

Ostensibly, shifting from riparian to prior appropriation rights is accomplished by a stroke of the síndico's pen; he declares the municipality water under rule (*en regla*). This action brings all irrigators under his authority by activating the previously described hierarchy (Figure 10.3). While water is under rule, the síndico maintains a water distribution list. Sections (tramos) are apportioned water on a "first come – first served" basis. Each receives a fixed number of hours or impoundments, varying with the type of appropriative methods the síndico feels most equitable. After each irrigation, a section may put its name on the list for an additional turn.

This period of prior appropriation ends when either (1) the river goes dry or, less commonly, (2) the first floodwater crashes down the escarpment – whichever occurs first. Thereafter, the riparian doctrine prevails until the síndico again declares water "under rule."

Timing of these shifts from one doctrine to the other would appear to confirm the excess-scarcity hypothesis used to explain the geographical distribution of the doctrines in the United States. However, the correlation between a physical fact (water quantity) and a cultural fact (the allocation system) demands the interlinking of variables explicable at the level of observable social behavior.

How does the switch from riparian to appropriative rights occur? Closer examination reveals that the síndico's activation of the prior appropriation doctrine rests, in part, upon a social-psychological mechanism: an individual's tolerance to stress. During the riparian period while rainfall and stream flow are high, conflicts over water rights are low. As precipitation and stream flow decrease, the frequency and ardor of conflicts over water rise. These conflicts are taken directly to the síndico, because the administrative water hierarchy is inoperative. He resolves these disputes in his capacity as an authority in the local level political system, not as an irrigation system administrator. Finally, the síndico grows weary and overtaxed with water disputes. Without warning, he declares the system under rule. This means that no one, including the síndico, can anticipate when prior appropriation begins.

Once prior appropriation is established, conflicts between irrigators subside. This decrease may be attributed to structural change in the adjudicative processes. A single individual, the síndico, mediates disputes under riparian rules. Under appropriation, a three-level hierarchy of mediators buffers disputes before they reach the síndico. If those conflicts unresolved by the lower level in the hierarchy reach the síndico, they are considered much more serious by the officials and community than disputes occurring under riparian periods. The former conflicts are those that could not be resolved by peers in the fields or by the tramo officials. The síndico now has public support to exercise his most coercive power – shifting the dispute to district court under the guise of a "civil disorder." Thus, the ultimate power for water regulation lies outside the local level. This power is based upon the síndico's dual role as local-level water administrator and civil official in the Mexican state.

Conflicts between sections (tramos) are rare because the criteria for membership prevents sections from becoming corporate groups. Partible inheritance scatters most households' fields, meaning that many households are members of several sections. Just as a kindred system creates cross-cutting links between individual kinsmen, making corporate kinship alliances logically impossible, so multiple section membership creates conflicts of interests between members. Informants provided this explanation when asked why no inter-section conflicts could be recalled.

This structure explains why conflicts subside under the prior appropriation doctrine, but fails to account for the flare-up of conflicts at the end of the rainy season. The excess-scarcity hypothesis also lacks correspondence with the data. If scarce precipitation and stream flow lead to frequent conflicts, then disputes should be high during the beginning of the rainy season when these same hydrologic conditions

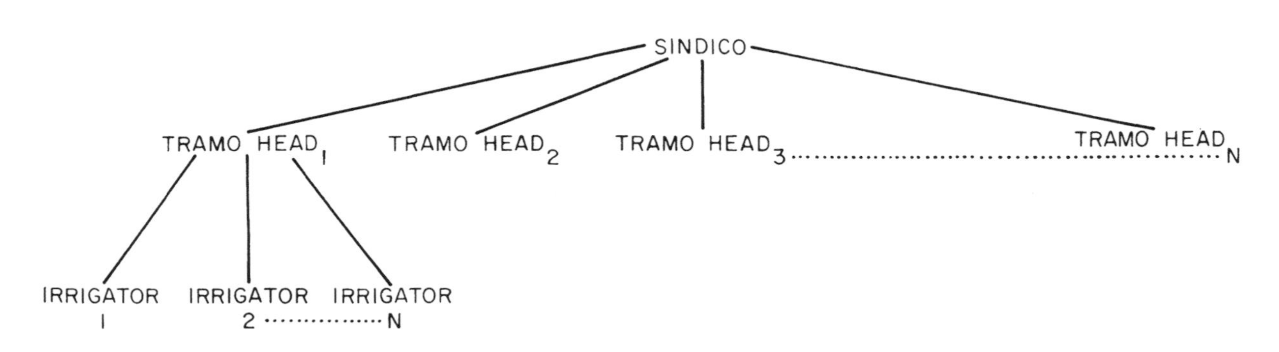


Fig. 10.3 Present social organization of water allocation (1970).

hold. That is, the "low precipitation = scarce water = more conflicts = more social control" explanation breaks down, because the system is under riparian rights at the beginning of rains, and under appropriation rights at their end, while precipitation is equally sparse at both extremes. The key to this puzzle lies in defining water scarcity in terms of the moisture needs of cultigens rather than in terms of absolute water quantity.

## MOISTURE-SENSITIVE PERIODS

In 1899, Brounov introduced the concept of a plant's critical period, meaning that part of a plant's developmental cycle most sensitive to meteorological events. Salter and Goode (1967) refined Brounov's work by introducing the concept of a plant's moisture-sensitive period (MSP), referring to that point or points in a plant's developmental cycle at which it is most sensitive to either excess or scarce moisture conditions. They review the literature for several hundred cultigens and demonstrate that these MSP's are crop specific.

Agronomists' research on the moisture-sensitive period of maize is of direct interest in this ethnographic case. The MSP of maize has been defined using observations of its differential yields under conditions of a) irrigation application experiments, b) soil drought response, and c) statistical studies of its response under different precipitation patterns. Using irrigation application experiments, Cordner (1942) discovered that irrigation before silking increases the plant's size, but not its yield; irrigation after this point brings about increasing yields. Howe and Rhoades (1955:98) found that "the yield of plants irrigated just before tasseling and during silking (144 bu./acre) was almost as high as, and not significantly different from, plants irrigated throughout the season (153 bu./acre)." Their control, unirrigated maize, yielded only 69 bu./acre, while a single irrigation in late tasseling yielded 101 bu./acre.

Robins and Domingo (1953) and Denmead and Shaw (1960) used an alternate strategy for determining maize's response to soil moisture. They allowed plants to deplete the moisture to a permanent wilting percentage at various stages of its development. They found that such deficient moisture conditions maintained for two to three days during tasseling and silking reduced yields 22 percent, while six to eight days of moisture deficiency reduced them 50 percent. Depletion of water during other phases of the plant's development failed to show such dramatic effects. Statistical studies of rainfall's effect upon maize production corroborated these experimental findings.

Salter and Goode (1967:32) summarize over twenty-five such controlled experiments:

> The results of both experimental and statistical studies thus suggest that soil mositure conditions during the period of flowering and early grain formation are particularly critical in determining grain yield in maize. They are reflected in many recommendations made over a number of years to irrigate the crop during this period.

## IRRIGATION'S MULTIPLE FUNCTIONS

This information on the moisture-sensitive period of maize permits further specification of water's place in Díaz Ordaz agriculture (Figure 10.4). Floodwater irrigation lengthens the annual growing season by placing a small, yet crucial, amount of water on the fields four to eight weeks ahead of the onset of rainfall adequate for dry farming. Given a four-month growing season for maize, the MSP for the first irrigated crop falls during the peak of the rainy season. This crop's irrigation needs are minimal, and rainfall is sufficient. However, limited irrigation does occur during this period. The variation in rainfall distribution leaves some fields drier than farmers feel necessary, and they may irrigate (under riparian rights). However, such irrigation involves risks. If a heavy downpour succeeds the irrigation, excessive saturation of the fields can cause erosion and crop damage. Due to these uncertainties, most informants choose not to irrigate.

The extended growing season also makes possible a second crop; its planting *immediately* follows the first harvest. Adequate rainfall during the planting period assures the second crop's germination. Heavy rainfall during the pre-tasseling period produces substantial plant foliage. Indirectly, the correspondence of heavy rainfall with early plant growth stimulates positive feedback to agricultural production. The leafage provides draft oxen with fodder during the dry winter months. Therefore, the well-fed oxen are capable of constant harrowing during the period before spring planting. Some farmers consider the second crop a "moderate success" if it yields only fodder for the winter.

The second crop's moisture-sensitive period occurs during the time of decreasing rainfall (about late October), and creates problems distinct from those for the earlier crop. Irrigation during this MSP differentiates those farmers harvesting only fodder from those harvesting fodder with grain. Unlike the situation in other periods, water now becomes a factor limiting productivity. Provided that the earlier crop meets a farmer's subsistence needs, the second crop is converted into the few amenities enjoyed by subsistence farmers.

Given the uncertainties of receiving rainfall and irrigation water during this period, and potential dangers from the onset of frost, some farmers opt for less moisture-sensitive and frost-resistant crops like alfalfa, chick peas, and wheat. These command a lower market price, but are less risky if water proves scarce.

With these physical constraints, the increase in water disputes during the second crop's moisture-sensitive period seems understandable. Water equals grain and Zapotecs know it. In turn, the low frequency of water-related disputes during the beginning of the rainy season seems reasonable. Water would be scarce during the earlier MSP if fodder were needed (i.e., to increase the plant's foliage), but alternative sources of fodder reduce this necessity.

The high frequency of disputes seems explicable given this modification of the water-scarcity hypothesis to take into account the specific water needs of maize. But the cultural response (the specific manner of water allocation) cannot be so easily linked to hydrologic conditions.

Activation of the prior appropriation doctrine and the administrative hierarchy echoes a theme of equality common to other sectors of Zapotec culture (for example, in equal inheritance rights of all heirs and equal civil obligations of family heads). The overall effect of the water distribution system prevents an individual from gaining an advantage over his neighbors. Likewise, the administrative structure with its rotation of officials makes it impossible for any one individual to take advantage of his administrative authority for any length of time. Thus, the structure of the administrative system proves consistent with the method of allocation itself.[6]

## IMPLICATIONS

This analysis has attempted to clarify the key characteristics of an irrigation system, but the painfully small sample impairs generalization. Nevertheless, most case studies implicitly hypothesize some model, and my discussion is no exception. I suggest that water scarcity must be defined in terms of the availability of water to plants during their moisture-sensitive periods. In drier climates, irrigation proves

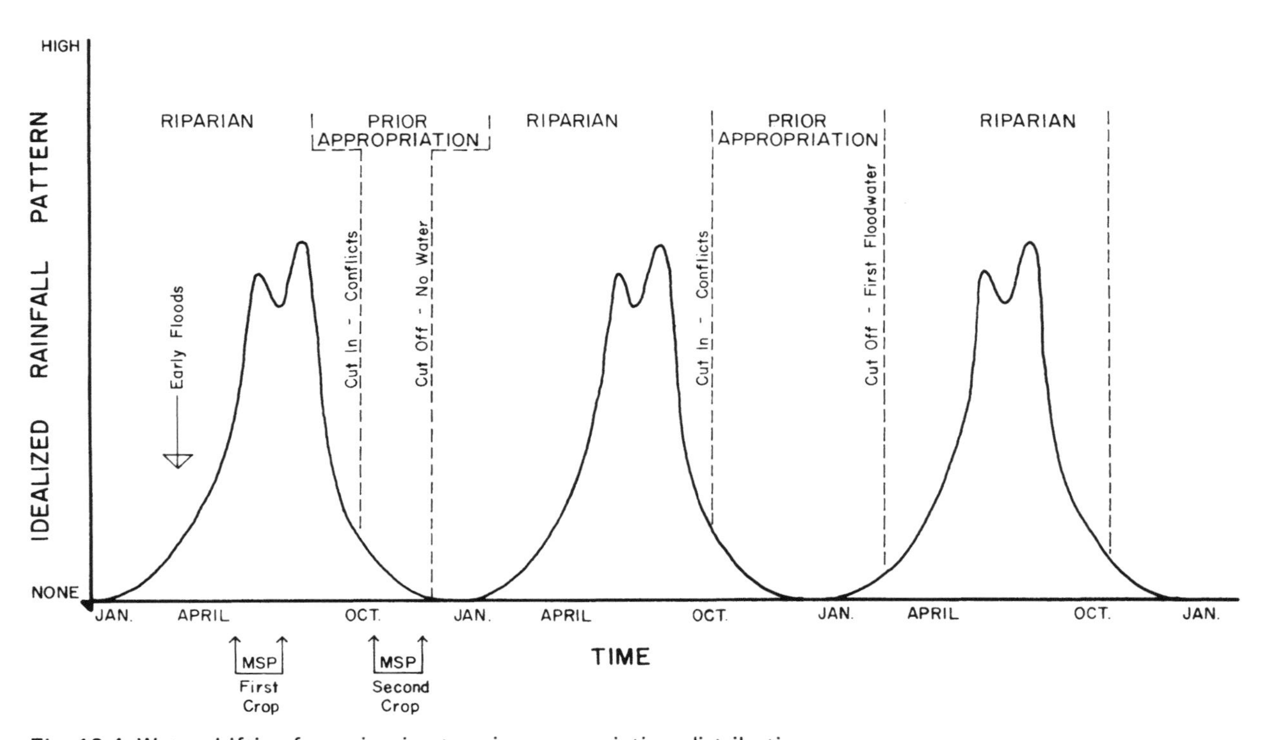


Fig. 10.4 Water shifting from riparian to prior appropriation distribution.

either an alternative or a supplement to rainfall. In determining the water scarcity of a system, the relative importance of these two sources must be considered. Hopefully, some day, rainfall and stream flow models from hydrology will be adapted to a discussion of this anthropological problem.

Assuming climatic conditions in which the entire planting of a society's main crop must take place at approximately the same time, I hypothesize the water scarcity during the moisture-sensitive period forces some kind of rigid social organization to allocate water. Whether the allocation favors equal or discriminatory distribution depends upon cultural factors, not water availability.

This conclusion reflects that of others in this volume. Irrigation is a phenomenon more complex than such crude nursery facts as that water evaporates and flows downhill. The failure of anthropological theory to link types of social organization with irrigation systems may be traced to its current inability to refine both sides of the irrigation-society equation. Although anthropologists arduously typologize and theorize about society, their work on irrigation lacks this maturity. Until irrigation systems are as adequately discriminated as social systems, theories relating them remain as trustworthy as a surgeon with poor eyesight.

## NOTES

[1] I wish to acknowledge the invaluable assistance of Cecil Welte, Director of the Oficina de Estudios de Humanidad del Valle de Oaxaca, who provided the rainfall data for the Tlacolula wing used in Figure 10.1, and to McGuire Gibson, Wayne Kappel, Chester Kisiel, and Dianne Wightman for their comments and revisions of earlier versions of this paper.

[2] In addition to floodwater and clear water, Díaz Ordaz Zapotecs exploit groundwater, using a recent technological innovation which they call an "underground dam" (*presa submerida*). This device is an unusual variation of a *kanat*. For a description of a real kanat, see Neely's paper in this volume. Informants claim this device was introduced in 1932 by the well-known Mexican engineer, Jorge Tamayo.

[3] No Spanish word for "canal" is used in this village except by a few village farmers who worked as *braceros* in the United States.

[4] This paragraph refers to a canal's longitudinal extension. Canals are moved several meters latitudinally up or downhill on a "trial and error" basis until the appropriate topographic location is reached. This relocation was still occurring during my field work in 1970.

[5] This measure roughly approximates field size (Downing, 1973).

[6] Explication of this theme of equality falls outside the scope of this paper, but is critical to any explanation of Mesoamerican social structure.

## REFERENCES

BROUNOV, R.I.
1899 The Dependency of Grain Yield on Sun Spots and Meteorological Factors. (Russian) Met. Bjuro uc Kom. S. Peterburg, Reprinted in P. I. Brounov Izbrannye Socinenija, Vol. 2, Sel'skohozjajstvennaja Meteorologija, Gidrometeoizdat., Leningrad. 57-67.

CORDNER, H. B.
1942 The Influence of Irrigation Water on the Yield and Quality of Sweet Corn and Tomatoes with Special Reference to the Time and Number of Applications. *Proceedings of the American Society of Horticultural Sciences.* 40:475-82.

DENMEAD, O. T. and R. H. Shaw
1960 The Effect of Soil Moisture Stress at Different Stages of Growth on the Development and Yield of Corn. *Agron. Jour.* 52:272-74.

DOWNING, Theodore E.
1973 Zapotec Inheritance. Ph.D. dissertation. Stanford University.

FLANNERY, Kent V. Anne V. T. Kirkby, Michael J. Kirkby, and Aubrey W. Williams, Jr.
1967 Farming Systems and Political Growth in Ancient Oaxaca. *Science.* 158:445-53.

HOWE, O. W. and H. F. Rhoades

1955 Irrigation Practice for Corn Production in Relation to Stage of Development. *Proceedings of Soil Science Society of America.* 19:94-98.

KAZMANN, Raphael G.

1965 *Modern Hydrology.* New York: Harper and Row.

LEES, Susan

1973 Socio Political Aspects of Canal Irrigation in the Valley of Oaxaca. Ann Arbor: Memoirs of the Museum of Anthropology, University of Michigan.

PARSONS, Elise Clews

1936 *Mitla: Town of Souls.* Chicago: Univ. of Chicago Press.

ROBINS, J. S. and C. E. Domingo

1953 Some Effects of Severe Moisture Deficits at Specific Growth Stages in Corn. *Agron. Jour.* 45:618-21.

SALTER, P. J. and J. E. Goode

1967 Crop Responses to Water at Different Stages of Growth. *Research Review* No. 2, Commonwealth Bureau of Horticulture and Plantation Crops, East Malling, Maidstone, Kent, England.

SCHMIEDER, Oscar

1930 *The Settlements of the Tzapotec and Mije Indians.* Berkeley: Univ. of California Publications in Geography.

CHAPTER 11

# THE STATE'S USE OF IRRIGATION IN CHANGING PEASANT SOCIETY

Susan H. Lees
*Department of Anthropology, Hunter College*

The primary concern of many archeologists and culture historians in regard to canal irrigation has been the interpretation and analysis of its role in the development of complex hierarchical social systems. The problem is one of systemic change, involving positive feedback, in which the result is not only a more complex and hierarchical system of control, but also an increased capacity to respond to extra-local stimuli or perturbations.[1]

By systemic change I mean alterations in the integration or cohesiveness of the parts or subsystems of state-organized societies. Such alterations may result in an increased capacity to exploit energy sources, hence in increased production for the system as a whole. My concern is not directly with increased production; rather, it is with the organization of the subsystems of a society, and the manner in which a change in this organization will result in an increased rate of response by the subsystems and the system as a whole, to stimuli or perturbations from non-local sources. A positive fedback chain of responses which involves increased production and flow of material, energy, and information in the system is another way of saying that "progress," in the colloquial sense, is taking place. In such a system, new technology is rapidly introduced, becomes diffused, and affects a great proportion of the population and its various activities. I will take the position that more cohesive systems are potentially unstable.[2] Furthermore, due to this instability, state-organized societies tend to experience cycles, in which they return, periodically, to a less cohesive condition, then again to a more cohesive one.[3] One reason, I will suggest, for this instability and cyclical return to a less integrated condition, is the reduction of sensitivity of more cohesive systems to micro-environmental or localized perturbations. Generalized progress involves greater sensitivity to other sorts of perturbations at the cost of localized desensitivity. Long-term evolutionary success entails a capacity on the part of systems to respond to perturbations of all sorts, including new or unprecedented ones. (Slobodkin 1968). Therefore, states which survive have the capacity to alter their organization periodically to respond to new perturbations, and later return, through a variety of homeostatic mechanisms, to a more stable condition. This cycle, of course, refers to the system of organization, not population or technology, which may have been irrevocably altered by organizational change.

My emphasis on this hypothetical cyclical process has several purposes. The phenomenon which I am about to describe is only a slice out of a larger process; similar occurrences of the phenomenon elsewhere or at other times are also periodic, not constant. Further, in like manner, the role of canal irrigation in the development of early states changed periodically. In other words, the function of canal irrigation as a technological device should not be viewed as constant, but as variable with time and situation.

The majority of states in human history have been agrarian; that is, they have been supported largely by a peasant base. Thus, my analysis of the subsystems of a peasant society should have fairly general applicability. My example is drawn from contemporary Mexican society, in particular the Valley of Oaxaca in the south central Mexican Highlands.[4] The type of local subsystem organization found here is fairly common throughout the world, as is indicated in the literature on peasant society, particularly writings by Eric Wolf (1955, 1957) on the closed corporate community. At a higher level of generality, my analysis should also apply to an even greater range of peasant-based societies, including those of the early Old World civilizations.

In order to consider the nature and consequences of systemic change, we will start with a description of the relatively stable conditions of the system prior to perturbation and response. A highly stable form of organization was instituted in Mexico by the Spanish

Colonial government in the late seventeenth century. It remained in this stable form, with a few minor interruptions, for nearly four hundred years.

This form of organization was hierarchical and, more important for this discussion, what I will call "nearly decomposable."[5] Basically, the hierarchy had two levels. At the upper level was the state government, whose personnel were a small group of full-time governmental specialists. This group regulated interaction between the various subsystems at the lower levels, and more importantly, between the entire system and its external social environment – foreign relations.

After the instituting of a generalized form of organization for the subsystems at the lower level, the state government played a relatively insignificant part in their internal operation. Between lower-level subsystems there existed regulatory networks of interaction, perhaps the most important of which was the local market system. The market system and other networks of interaction were decentralized, as were the local systems of social control and resource exploitation.

Thus, local units of the population responded primarily to micro-environmental conditions and perturbations, each adjusting itself more or less independently from the whole. Although the Mexican Revolution in the early part of this century had considerable consequences for local organization and distribution of land and resources among local populations in some parts of the country, significant lag in other areas – like Oaxaca – in social reform, bears witness to the basic decentralization of local subsystems in Mexico, and to the pronounced stability of the larger system as instituted by the long-extinct Spanish colonial government.

In the political sphere, perhaps the most striking feature of local subsystem organization was the so-called closed corporate peasant community. Each peasant community had a relatively autonomous local government, different in form and mode of operation from the state government. At the local level, officials were selected from the community at large to serve as organizers of collective activities and keepers of the peace. Official positions were circulated among all adult male community members so that individuals could specialize neither in any particular official function, nor in the full-time occupation of social control itself, as was the case in the upper level of the political hierarchy. The task of social control were firmly divided between the upper and lower level subsystems.

Irrigation resources in the Valley of Oaxaca derive mainly from small streams originating in the mountains and piedmont to the east and west. Since the Revolution, these resources have been the exclusive property of local communities; before the Revolution, some were controlled by local private entrepeneurs. Control of access to, and maintenance of irrigation facilities was in the hands of official delegates of the local communities. The form of control, in terms of the type of officials in charge and the rules pertaining to access, varied widely from community to community in response to local environmental and historical conditions. Inter-community interaction in regard to shared use of a single water source was also locally controlled, except in rare cases of violent dispute, which even then almost never involved state intervention.

There was little variation in irrigation technology between localities; that which existed was, again, the result of local response to microenvironmental and historical conditions. Adjustments to variations in the availability of water were made on a community or even on an individual basis.

Local peasant communities tended to be highly egalitarian. The rule of continual circulation of political offices among community members maintained this egalitarianism in theory, and to a considerable extent in practice. Other social mechanisms – homeostatic devices, if you will – also functioned to some extent to level differences in wealth, hence to restore the basis for egalitarian participation in community affairs.

Local government worked primarily on a consensus basis; hence innovations were accepted rather slowly. There was considerable resistance to change which would alter the egalitarian basis for the functioning of the community system of social control.

Among other factors, the combination of wealth-levelling devices, a consensus system of government, and social inhibition of competition tended to lower the effectiveness of resource exploitation, and consequently, the general level of agricultural production. Wolf (1955) and others have remarked that such communities tend to be relatively poor, to exploit their resources at less than maximum capacity, and to operate close to the subsistence level. It can be demonstrated that a return to community control of certain water resources since the Revolution has resulted in a decreased effectiveness of use, a general decrease in the productivity of such resource exploitation (Lees 1973). Although new technology might help to increase effectiveness of water resource exploitation, technology is not the only factor; organizational variables also affect exploitative capacity. In this case, less than maximal resource exploitation and productivity can be seen as a partial consequence – or perhaps an integral aspect – of a decentralized, isolated, egalitarian political subsystem within the larger state organization.

Now, let us turn to the process of change. At present, the system, the state of Mexico, is experiencing perturbations of various sorts, two of which I shall briefly refer to. The source of one is internal, and the other, external.

The source of internal perturbation is an upset in the balance of population, land, and water. Population is on the increase, particularly in the densely settled urban centers, and agricultural production has not kept pace with this rise in population. Arable land is in short supply. Increasing the agricultural productivity of the existing arable land calls for greater use of water resources. At the same time, the growth of urban centers and their consequent population concentration results in a water shortage. Mexico's problem, then, is to scale the availability of water resources to increased demands for agricultural production, while also providing water for its cities and industries.

The second source of perturbation arises from Mexico's relations with other nations. Mexico is tied into a world market, and is presently in a disadvantageous position: it produces too little for trade, and is required to buy too much. In order to improve its position, it must increase production in various sectors of the economy. Since its economy rests on an agrarian base, it must, at least for the present, increase its agricultural capacity. As I indicated above, this improvement would require an expanded use of its irrigation resources. At the same time, Mexico is attempting to compete industrially; this competition requires power, which, at present, means harnessing and using more water sources.

These two sources of perturbation, then, reenforce one another, to place pressure upon the state water sources and the technology and organization applicable to exploitation of water sources.

The function of the state government subsystem, as I explained above, is to regulate interactions between other, lower level subsystems, and interactions between the whole system and other, external state systems. Therefore, it is this high level subsystem which eventually experiences the perturbations, responds to them, and attempts to correct the disturbances. One of its most important responses is the establishment of a centralized program for hydraulic development.

The Mexican government's hydraulic development program rests, in part, upon a policy of installing numerous small-scale irrigation devices.[6] These devices are relatively inexpensive and technologically simple. The government itself provides engineers, equipment for construction, some materials, and some financing; the local community provides some labor and financing. The government agents decide the installation's location, construction, and use. The government demands that a sector of the community – those who will be using the irrigation facility – select from among themselves a committee responsible for carrying out rules and directives pertaining to access, maintenance, and allocation of water among themselves. These officials are not responsible to the local governing body, but to agents of the central state government. A new form of local control is instituted, one which is not autonomous, and which responds, in part, to extra-local conditions. Thus, there is an organizational change within the system as a byproduct of technological change.

By installing new irrigation devices whose construction and maintenance is beyond the capacity of the local peasant community, the state government both reduces local village autonomy and causes the local community to become responsive to non-local perturbations. One area of response is increased exploitation of water resources and, at least potentially, an increased level of production; this is a response to pressures that are non-local in origin. The results of such a response may be highly disruptive to the formerly stable local government system. Homeostatic devices for maintaining equality and noncompetition among community members may no longer function, particularly in the face of reduced authority of the internal community government, once external officials from the state government have intervened as higher authorities for adjudicating disputes and appointing water-control officials. Communities are given a greater capacity to produce through technological innovation, while at the same time, social conditions which formerly inhibited productivity are disrupted, further enabling or inducing the community to meet external pressures to increase productivity.

A second area of the hydrological development program, not so explicitly stated as policy, is the assertion of authority by the state government over the allocation of water resources between communities and other users. This assertion of authority is part of an attempt to reallocate water resources in such a way as to respond to pressures of non-local origin.

For example, as I mentioned above, industrial development is one area of the state's scheme of response to external pressures. A factory producing textiles for sale abroad may require water power; the centralized irrigation authority may allocate water to the factory, depriving local farmers of their traditional irrigation source.[7] Here is one example of insensitivity to local conditions. The needs of the larger system may require an immediate response in the industrial sector of the economy; hence, the act of allocating water to the factory may benefit the larger system, though its immediate consequences for the local agriculturalists may be disastrous. On the other hand, local farmers may ultimately benefit by such an act: the factory may provide employment for the farmers' sons. That employment may in turn provide funds for the purchase of tractors and mechanical water pumps which will permit the local farmers to bypass traditional agricultural resources (grazing land for oxen, canal water), and at the same time, increase production in the agricultural sector. The ultimate response, then, would be what might be called "progress."

Government intervention has sometimes changed inter-village relationships. In certain cases, a more equitable distribution of water between upstream and downstream villages has resulted. Certain upstream villages which once sold water to downstream villages are now deprived of that source of income. Other villages which once took all the available water for themselves, must now share water with others. The amount they receive is, in part, proportional to their size. Thus, the amount of water available to such a village depends upon the number of irrigators in another community using the same government-installed irrigation facility. Local geographic and historical conditions are ignored; a village with a long history of dominance over others in its region due to its monopoly over the irrigation resource becomes their equal.

Decisions about the allocation of water may be affected by politics at the higher level. One village may be favored over another for political reasons. Thus, once isolated communities are drawn into national politics and are forced to respond to perturbations whose source, once again, is elsewhere.

Government intervention at the local level may have rapid and significant consequences for local community agricultural output. New irrigation devices have two contrasting results. On the one hand, they may result in immediately increased water availability and a sudden increase in population. On the other hand there may be a sudden decrease in water availability, due to over-tapping of the water table. Here is another example of insensitivity to local conditions. A local response appropriate to over-use of the water supply would be to cut back water use until the resource had replenished itself. But the larger system demands or results in a different response – hydraulic facilities are increased in size and number, further damaging the water supply. The response of the larger system is to grow and expand, not to cut back. This expansion may have far-reaching consequences for villagers of the region, who may be forced to find employment in the cities, ultimately placing greater pressure on water resources and the already limited arable land to produce enough to support its urban populace. This local insensitivity starts a positive feed-back chain whose consequences are felt more rapidly by the now more integrated system, which is called upon to speed up still further responses to now increased pressures. In other words, a response which began as a conscious and purposive effort to correct a deficiency in the system becomes itself the source of further disturbances which demand increasingly rapid and greater responses. This positive feed-back chain introduces increasing instability into the system, and reduces its capacity to restore itself to a former state.

To summarize, by installing irrigation devices and asserting authority over the allocation of water resources, the central government subsystem mobilizes the greater system to respond as a whole, and among its parts, to external and internal perturbations. In doing so, the system changes more rapidly, evolves, and becomes more unstable. Eventually, a new equilibrium may be reached, and the system may begin to respond once again to local conditions. This process entails a breakdown of links between the various subsystems, a greater decomposability, hence a greater stability.

At certain stages of their development, sub-units of the populations of regions occupied by peasant agriculturalists are relatively autonomous and decentralized, linked together by loose networks of political, social, and economic interaction, such as local market systems and weakly assertive political confederations. Perturbations arising periodically from external competition and trade or from internal reorganizations, or both, place pressures upon the system at large, and particularly upon political subsystems whose function it is to regulate internal and external interactions. These pressures are, at least in part, to increase agricultural productivity. In those areas in which agricultural productive capacity depends to a great extent upon water control, irrigation resources and facilities provide a primary locus of response through centralization and mobilization. Intervention in the construction, maintenance, or control of local irrigation systems brings about a decrease in local autonomy, and a concommitant increase in agricultural productivity. In the process, local conditions are ignored, bypassed, or disregarded in favor of response to non-local perturbations. The result is comparatively rapid change and the evolution of new forms of response, part of which are technological, and part sociopolitical. Changes are diffused more rapidly throughout the system as it becomes increasingly responsive and cohesive. Such cohesiveness or hyper-dependence makes such systems unstable; they change and evolve, but at the same time become vulnerable to breakdown.

## NOTES

[1] A debt of gratitude is owed to Daniel G. Bates of Hunter College, CUNY, who read this paper and made helpful comments, and to members of Professor Vayda's "Wednesday night" seminar on Human Ecology at Columbia University. Their stimulating discussions inspired this paper.

[2] See Rappaport (1969) for a discussion of "hypercoherence" and systemic instability.

[3] Gibson (this volume) suggests a cyclical pattern for early Mesopotamian civilization; Wittfogel (1957) and Lattimore (1940) described cyclical patterns for ancient Chinese civilization; according to my interpretation, pre-Hispanic Mesoamerican civilization also followed such a pattern.

[4] See Lees (1973) for a more complete description.

[5] Simon (1962) uses this term to characterize hierarchical systems.

[6] For description and policy statements, see *Boletin Informativo de la Direccion General de Ingeniería Agricola* Nos. 1, 2 and 3, published by the Secretaria de Agricultura y Ganaderia of Mexico.

[7] This was the case, for example, in the village of San Agustin Etla, where the textile plant, Fabrica San José, was constructed. For documentation of this case and others which follow in the text, see Lees (1973).

## REFERENCES

DIRECCION GENERAL DE INGENIERÍA AGRICOLA
1968 *Boletin Informativo de la Direccion General de Ingeniería Agricola, Nums. 1, 2, and 3.* Mexico: Secretaria de Agricultura y Ganaderia. July, September 1968, January 1969.

LATTIMORE, Owen
1940 *Inner Asian Frontiers of China.* New York: American Geographical Society of New York.

LEES, Susan H.
1973 Sociopolitical Aspects of Canal Irrigation in the Valley of Oaxaca. Vol. 2 of the *Oaxaca Reports;* Memoir 6 of the Univ. of Michigan Museum of Anthropology. Ann Arbor: Univ. of Michigan Museum of Anthropology.

RAPPAPORT, Roy A.
1969 *Sanctity and Adaption.* Prepared for Wenner-Gren Symposium "The Moral and Esthetic Structure of Human Adaptation." New York: Wenner-Gren Foundation.

SIMON, Herbert A.
1962 The Architecture of Complexity. *Proceedings of the American Philosophical Society.* 106:467-82.

SLOBODKIN, L. B.
1968 Toward a Predictive Theory of Evolution. In *Population Biology and Evolution.* Syracuse: Syracuse Univ. Press.

WITTFOGEL, K.
1957 *Oriental Despotism.* New Haven: Yale Univ. Press.

WOLF, E. R.

1955 Types of Latin American Peasantry: A Preliminary Discussion. *American Anthropologist*. 57:452-71.

1957 Closed Corporate Communities in Mesoamerica and Central Java. *Southwestern Journal of Anthropology*. 13:1-18.

CHAPTER 12

# IRRIGATION, CONFLICT, AND POLITICS
## A Mexican Case

Eva Hunt
*Department of Anthropology*
*Boston University*

and

Robert C. Hunt
*Department of Anthropology*
*Brandeis University*

### INTRODUCTION[1]

One of the most controversial sets of functional hypotheses about relationships between human social institutions has to do with the nature and consequences of irrigated agriculture. Now primarily associated with the name of Wittfogel (and the society type he called "Oriental Despotism"), there have been several other scholars of note who have worked this particular mine (cf. Steward 1955; Wittfogel 1957; Millon 1962). Price, in a recent article (1971), has stated the hypothesized relationships between systems very clearly and succinctly. She quite properly separates them into two parts. The first is a set of statements about the cause and effect relationship between the invention of irrigated agriculture and the origin of the state. As she clearly and effectively argues, this problem is diachronic, and there are only a half dozen or so cases which are clearly relevant to the discussion. We shall not be concerned with this part of the problem here.

The other half of this endeavor concerns statements about the synchronic functional relationship between irrigation, on the one hand, and various parts of the socio-cultural system, on the other. Mainly, these propositions have to do with the domain of the social system we usually call *political institutions* or *political structure.* It is to this nexus of functional statements that we wish to address ourselves.

There is a single major model in the literature of the relation between the two systems. Wittfogel and his followers, using Marx's ideas about Asiatic society, have characterized the prototypic irrigated society on the basis of their model of China and southeast Asia. In such societies, the political life is controlled by an elitist bureaucracy of full-time specialists with both secular and religious (theocratic) powers. This bureaucracy also has a monopoly of many other social institutions, such as entrepreneurial activities and military organizations and action. It also has close to total control of the economic system, appropriating for itself all privileges derived from the control of the economic surplus created by the masses of producers. The ruling elite is not simply powerful, but is also highly centralized, small in numbers, physically concentrated, and despotic. Wittfogel argues that such despotic governments emerge in pre-industrial societies which have an agricultural base supported by irrigation. The basic need to control water, administer the construction and maintenance of massive water works (by corvee labor), allocate capital expenditures and resources over water, as well as to schedule the distribution to keep the system going efficiently, gives those persons in the irrigated society who occupy crucial role synapses in the irrigation system, expansive powers in other domains of social life. In a sense, the ruling elites are supported in their increasing political control by the inherent inter- and intra-societal competition for the artificial control of water, which leads to a further need to enforce peace from above. On the other hand, high centralization and concentration of power facilitates the cooperation of larger masses of people for the common goals of the irrigation society, which frequently require large labor inputs. The ruling bureaucracy builds and reinforces its power due to the fact it can effectively undertake the functions and roles necessary in management which the masses cannot efficiently muster on the basis of consensus. The bureaucracy's monopoly, thus, is self-reinforcing, since they can use water to control the society, sanction revolt against their power, and build power in other realms of social life, which then again become self-reinforcing and self-serving.

This set of propositions deals with a) the nature of total political power which a society practicing irrigation can generate; b) the degree of cohesiveness the political elites can command; c) the degree of concentration of political power in a few privileged personnel (a bureaucracy, a limited set of governmental roles, etc.); d) the degree to which political power generated in the realm of irrigation can be generalized or transferred to other domains of the society; and e) the degree to which irrigation generates both cooperation and competition within a society, and between societies.

Wittfogel has modified his prototype hydraulic society by indicating that some of the conditions of his ideal model are not likely to be met when the scale of the society, the scale of the irrigation system, or other historical conditions are not met. However, part of his major theory (i.e. the despotic quality of the state) applies only to a special case, that of oriental despotism, the major type being Chinese hydraulic society. The core elements of this special case are a "single" physical irrigation system, massive in scale, managed by a highly centralized despotic regime. Wittfogel has left himself open to very serious criticism because he has been less than clear, however, in operationalizing the variable of scale, and defining the ethnographic limits to which the core theory of despotism applies. Even in his most recent statement (1972) Wittfogel appears to include a vast range of societies which practice irrigation, as fitting his model, although it is obviously absurd to conceive of them as despotic political systems (e.g. the Hopi). The domain of societies which manage water for agricultural purposes ranges from primitive acephalous cultivators (e.g. the Nuer), to chiefdoms (e.g. Hawaii), to agrarian states (some of which had and others didn't have despotic regimes historically), to industrial states. Such variants only take one dimension (type of political system).

There is as yet no general model presented in the literature which can deal with all societies practicing irrigation. Nor is there a model which deals exclusively with some operationalized sample of small-scale systems and their attributes as opposed to those of large scale (whatever one may clearly define by small or large scale). Moreover, there has been in the literature an apparent confusion between political power and despotic political power.

Criticisms of Wittfogel's *model,* as opposed to evaluations of the accuracy of his assessment of particular empirical cases, must stay within the domain that Wittfogel has defined. Many such criticisms have been focused *not* on the possible relationship of irrigation, the economy, stratification, and politics, but on the supposed nexus of irrigation (usually generally conceived, rather than with the specific limitations demanded by Wittfogel's theory), centralization, and despotism. This distinction must be kept in mind as one reads the irrigation literature, for a major effort has taken place to discredit the despotism part of the theory. In order to accomplish this effort, the relationship between irrigation and the political structure has been obscured, which is a case of throwing away the grain to rid oneself of the chaff.

Robert Adams (1966:72-74) in the context of discussing the origin of the state, has been careful to separate other major effects of irrigation as a social force, from the variable of despotism of a managerial class per se. Adam's argument is clearly stated in the following quote:

> Our special concern here, however, is not with the importance of irrigation as one of the vital, interacting factors increasing agricultural productivity and hence surpluses, or even with its possible contribution to the growth of social stratification through its encouragement of differential yields. If perhaps not quite to the same degree, these effects of irrigation almost certainly were present in central Mexico just as they were in Mesopotamia. But it is Wittfogel's contention that the primary significance of irrigation arose not from its encouragement of new economic resources and social complexity but rather from the impetus given to the formation of coercive political institutions by the managerial requirements of large-scale canal systems. And in this respect the available evidence, closely paralleling that for early Mesopotamia, fails to support him.

Both Adams and Glick (1970) have investigated irrigation systems in a diachronic framework. They both reject the theory of a managerial, bureaucratic despotism, but find plenty of evidence for interrelationships of the irrigation system, production, role complexity, stratification, and the political system. Although Adams fundamentally disagrees with the notion that irrigation gives impetus to the formation of cohesive political systems, he doesn't reject, as Millon (1962) appears to have done, the implicit relation between irrigation and political processes leading to changes of social structure, or, more specifically, with irrigation system processes affecting the socio-political structure of population aggregates under a single political hierarchy. Price (1971) apparently misses the subtlety of this position.

Glick's evidence doesn't support the view of the existence of a despotic system in medieval Valencia. It presents good historical evidence, however, that the social consequences of irrigation were many and far-reaching, in terms of proliferation of political roles dealing with water control, adjudication, etc., and in terms of the generation and resolution of social conflict, harvest yields, and the historical build up of a complex political code of norms, laws, and traditions dealing with irrigation, of a large irrigated territory. Our own historical evidence for the sixteenth-century Cuicatec (briefly outlined here, but dealt with in great detail in E. Hunt [1972]) supports the view that there is no simple relation between despotism and irrigation (at least for small-size, closed, physical systems), but there is a clear relation between water control and the sources of power of the ruling elites of the irrigated society.

There have been many responses to Wittfogel's work; these responses have varied greatly in degree of objectivity, strength of evidence, and nature and kind of ideological commitment. The basic set of diachronic propositions, however, has not basically been expanded from Wittfogel's original statement.[2]

There is as yet no general synchronic model of the kinds of functional relationships between irrigation and other socio-cultural phenomena. Various studies have argued bits and pieces of the relationships. Leach (1961), for example, in his study of a village in Ceylon, argues that social relationships in the village are strongly determined by property, which is in turn determined by the water distribution system. Gray (1963), Fernea (1970), and Sahlins (1962) have all presented information on both the water management system and on the system of political roles, paying some attention to the relationships between them.

Millon (1962) has provided the only even partially systematic comparative attempt, in print, to investigate the synchronic relationship between the degree of centralization and the size of the irrigation system. He states that he is not concerned with Wittfogel's "hypotheses," and he uses what he calls "small-scale" systems. However, his variables are the same as those proposed by Wittfogel, as is the major problem. Millon concludes that in small-scale systems not only is there no necessary causal relationship between irrigation and centralization, but in fact that there appears to be no relationship at all! We have argued elsewhere (E. Hunt 1972) that Millon's results are suspect due to the biases in the sample and the ambiguities of concept definition. However, what is important here is that the literature presents us with two extreme, opposed positions, and a number of case studies in the middle, the results of which are extremely ambiguous insofar as they support either set of positive or negative hypotheses.

As a consequence of our symposium meeting at the 1972 Annual Meeting of the SWAA at Long Beach in April, it became clear that there are a series of unsolved methodological problems which have produced, in part, the inconclusive and contradictory results we are faced with. First, it appears that there is no such thing as an irrigation society. There are many varieties of society practicing irrigation, and many forms of irrigation. Specification of variables and the conditions under which they function has not been systematic. Large water works to control major floods may be greatly different in their effect on society from flimsy water works on small streams, or from the building of one isolated qanat. A situation in which upstream irrigators can control water flow has to be quite different from one in which they cannot. Paddy rice irrigation water moving through fields in countries with sufficient rain probably cannot affect social relations in the same manner in which maize irrigation water absorbed by a plot in an arid environment can. Moreover, we have all been remiss in specifying the meaning of our concepts. Such terms as *centralization, traditional practices, functional efficiency, small-* and *large-scale,* which are the cornerstones of the literature, have been left ambiguously defined at best, and at worst are only privately intuitive in meaning.

We are faced, therefore, with at least three major tasks. First, we must define our terms. Second, we must specify the range of subtypes of irrigated societies, in each case specifying the relevant range of variation of each major theoretically significant variable. (The theoretically significant variables must be specified in terms of a particular theory.) Third, we should test the propositions about relationships between variables in the only valid way, with a large-sample comparative study. We could ideally control for the range of variation of all variables involved, and the range of types of irrigation societies which will eventually emerge. Cross-cultural comparisons, however, require the existence of prior meticulous historical as well as synchronic analysis of specific cases. Such cases, which can be limited in number to cover each major irrigation system type (e.g. tanks, canals, chain wells, etc., of differing size and specified range) will provide the empirical basis for the necessary comparisons leading to less flimsy generalizations than we had up to the present time.

The successful application of the comparative method assumes the existence of a theoretical model which will cover most, if not all, of the variation found in the total population. This in turn implies a sophisticated knowledge of *each* of the types. These conditions can hardly be said to exist today with respect to the study of the social impact of irrigated agriculture. In summary, we need both good detailed case studies and comparative statements.

In this paper we propose to present a case study in depth, not a comparative study. We will present hypotheses and propositions which are restricted in application to a subtype of irrigated society. This subtype, the Cuicatec of Mexico, will be specified in detail in terms of its conditions, both ecological and socio-cultural. As such, this paper will first present relevant ethnographic data on the social impact of irrigation in one society. Second, it will attempt to clarify in part a few methodological muddles of conceptual definition. Third, it will attempt to specify in detail the nature of the variables. Specifically, we feel that for any system studied, the following variables require unambiguous treatment.

1. The natural conditions in which the system functions (e.g., climate, water balance, soil, topography, nature of water sources).

2. The physical irrigation system, including how water is extracted, moved, and stored. This description includes the scale of water works and the technology of irrigation from a mechanical point of view. Considerable attention should be paid to the maintenance needs of the physical system. This variable of maintenance has two dimensions. One refers to regular upkeep of a working system which is in balance (e.g. monthly cleanings and repairs); the other refers to major maintenance to cope with environmental degradation (e.g. increased levels of soil salinity, significant changes in the level of the water table). The Cuicatec case discussed here deals only with maintenance in the first meaning of the concept. The reason is simple: there is no evidence of basic environmental degradation for at least the four hundred-year period dealt with here. However, in adjacent areas north of the Cuicatec (the Tehuacan valley), maintenance in the second meaning is of major importance, since high levels of salt and travertine are present in the water, and a marked lowering of the water table has occurred in the last one hundred years. This lowering is partially related to enormous population growth and concomitant water demands in the valley.

3. The cultivation system. What crops are planted, their water demands, crucial watering periods (Downing's chapter in this volume indicates these are of major importance), ecological balance and competition between crops and between crops and wild plants; in other words, the general structure of the ecosystem within which domestic cultigens are produced.

4. The general characteristics of the cultural system, including the folk view of the environment and resources and the religious or ritually symbolic aspects which affect irrigation practices (usually including the calendar and round of sacred feasts and holidays dealing with water gods).

5. The general characteristics of the social system, particularly a description of the social structure and system of stratification as it affects the social aspects of irrigation, the institutions which deal with the management of public goals, conflict resolution, and allocation of resources; that is, the specific system of institutions which deal with politics, particularly "water" politics, and such other relevant aspects in a given situation as the impact of ethnicity on access to economic privileges.

6. A complete description of the system of roles and groups in the society which are directly linked and concerned with irrigation. Such a description should entail a discussion of to what degree and how such roles and/or groups are embedded in other parts of the social and cultural systems, or into roles or groups which have other primary functions.

In what follows, we have not been equally successful in dealing with each of these variables. Notably absent from this paper is an account of the cultivation system and the general cultural system. We have attempted, however, to utilize our data for all of these categories.[3]

In summary, we are operating in the context of a single synchronic case study, attempting to specify the conditions under which a particular system operates, and focusing on functional relationships which have already been proposed in the diachronic, historical literature with inconclusive results. Our purposes are to isolate parameters, specify a possible subtype of irrigation society, and generate hypotheses for future comparative testing.

Before we proceed any further, we must clarify the meaning of the concept of *centralization,* which although it has been a crucial variable in the anthropological study of irrigation, has been left defined at a level of ambiquity which makes it, at times, unusable. In our view, centralization of authority should be understood within two different contexts. One refers exclusively to authority in terms of the irrigation system. The other refers to generalized political authority, which may involve other functions of control outside or above simple water control. In one case, authority is exercised over different decision-making rights in terms, exclusively, of the social or technological needs of the irrigation system per se. In the other case, authority is exercised over water as one aspect of the decision-making rights of a complex political role or of a large, multi-function political machine.

It is usually the case in chiefdoms and state systems, that there is at least an incipient (if not well-developed) system of organic integration of the political community which involves multi-dimensional functions. These functions are allocated to territorial units (and the social groups they contain), which are hierarchically nested within each other. In this case, the degree of centralization could be measured by the extent to which a particular total function, such as water control, is located at the top level of the hierarchy, or at lower levels. In fact, however, it is most likely that different aspects of water control will be allocated at different levels of the total political structure. Higher levels may delegate some functions to local authorities, but with centralization *sovereignty* lies at the top levels of the political structure. Homologous (same size or same function) groupings do not make decisions about each other (for example, two villages of the same political status using the same stream), particularly if there is any question of competition, but decisions are likely to be made at the next higher level of the authority ladder or power hierarchy. When the latter is the case, decisions at the lower levels (e.g., decisions usually left to village officers) which appear as local in their power base, are not truly autonomous, because they are linked to a functioning central government structure which contains higher levels of decision making. These strata have the potential power to override lower-level decisions and introduce external controls if it appears necessary.

If we consider centralization of authority purely in the context of irrigation systems, the concept may refer to several different kinds of unified social control, i.e., to several separate, different variables. These variables can be located at different levels of the social structure. In addition to responsibility for construction and maintenance of a canal system (or dams, or water tanks), we must also include the rights to allocate water (one of Millon's major variables), the control and management of services connected with conflict resolution between water users, the duties connected with the legal or military defense of water sources and improved irrigated lands, the administration of labor resources to maintain the physical system, the responsibility for the proper performance of rituals insuring water supplies, and so on. These functions may involve a single role, or a set of specialized roles which may or may not be ranked vis-à-vis other political roles in the society. The functions may also involve a specialized institution, or simply be a set of functions embedded in other socio-political institutions expected to perform other functions.

Water control is linked to other social controls in more or less direct ways, but always in terms of the viability of the water-control mechanisms within a particular social structure. Societies which irrigate by canals have some clearly established networks of control in which differences in authority and differential power over water exists among persons, roles, and institutionally recognized social groups. Power differences about water, moreover, do not occur purely in isolation; water rights usually dovetail with the social allocation of other superordinate rights. Thus, wealth, social class, hereditary descent group membership in aristocratic lineages, and ascription to political roles which imply control of other resources may correlate with rights within the irrigation system. If centralization of authority is understood in these terms, allowing for the variations of a) degree of unification of decisions about several separate variables in a single role, set of roles, or institution, and b) intensity of embeddedness of the decision-making machinery about water in the larger political institutions, we have to talk about differences in degree of centralization rather than simple presence or absence. Moreover, measures of degree of centralization should involve several independent variables.

A good analysis will also focus on the real versus the ideal degree of local autonomy which exists in different systems. This focus is important because some systems (such as the one described in this paper and that of Pul Eliya (cf. Leach 1961; E. Hunt 1972) appear at first glance to be decentralized, due to smoothness of operation at the local level. Decisions appear to be made at this level when in fact for major variables they are not autonomous systems at all. Upon occasion the actors of the system may appear autonomous, and believe themselves to be so. In fact, roles are nested in a hierarchy of state-supported and legitimized water-control institutions, and are articulated in their actions into an elaborate system of linked activities. If roles and institutional activities are out of phase with each other, the harmonic workings of the traditional social arrangements concerning water are very likely to become disrupted.

Rather than an institutional form, centralized authority may refer to a set of roles which specialize in controlling, at the public domain level, water allocation, labor management, defensive warfare of water, internal conflict resolution for a particular irrigation system, and so on. For examples of roles centralizing arbitration and conflict resolution, one could mention the role of water judges in medieval Spain (Glick 1970).

In summary, the degree of total political centralization at the level of integration of the state (i.e., far above the limits of the irrigation system as a physical unit) in such cases as Iraq and Mexico, cannot be treated as equivalent to the degree of centralization at the level of a village in a tribal society or a district in a feudal one. We have asked: Can centralized authority be treated as an equivalent phenomenon in all these cases? Can the autonomous villages of the tribal Sonjo, a sub-tribe or a segment of a society dominated by a foreign metropolitan colonial power such as the British Empire, and a federalized national state such as Mexico be conceptualized as equivalent societies or even as having equivalent degrees of centralized authority? And if the answer is positive, what definition of centralization could make it possible? We believe they cannot, and levels of political integration and articulation need to be clearly differentiated (cf. Adams, Richard, 1971 for an illuminating theoretical discussion of this issue.) We also have to distinguish the anthropologist's and the native's view of the system. Irrigators may not be aware of the total system in which they participate, especially when a system is highly traditionalized. The natives may also not be aware of the social consequences of their system (Netting's paper shows this point clearly). Hence, the irrigators may see the system as egalitarian when in fact it makes for differential access to control of water supplies, or as decentralized when in fact it is not, simply because the potential of force inherent in the state safeguards the maintenance of order without constant supervision.

It would appear that the crucial factors in considering the issue of centralization are the degree to which roles are embedded, the number of roles that have control functions, and the degree to which the roles are connected with the apex of the political system. When roles are embedded, small in number, and connected with the apex, the system is highly centralized. By *embeddedness* we mean that the roles directly connected with irrigation are usually combined in a person(s) who control a number of other powerful roles integral to other parts of the economic and political institutions. By *number of roles,* we refer to whether or not a small number of roles are involved in directing or controlling the activities of a large number of persons. By *connectedness with the apex,* we mean the degree to which decisions made at any level of a system are responsible to, and respond to, actual or potential decisions made at the apex of the largest political system, including the social territory in question.

As we shall see in this paper, a major problem in understanding a state with irrigation is to determine how the local activities and structures are related to activities and structures in wider domains. For our purposes here, we shall refer to these two domains as local and national. A major point this paper makes concerns the articulation of the local and national decision-making agencies concerned, in one way or another, with irrigation.

A full understanding of this articulation demands a distinction between two types of task: frequent day-to-day decisions, mostly concentrated on maintenance and allocation; and infrequent decisions, usually linked with major construction, defense, and adjudication. The day-to-day decisions are manifestly under local control, sometimes involving specialized agencies with roles that are little embedded. The number of roles can be large or small, and there is little if any communication between these local agencies or role systems and those in wider domains (for these specific tasks). These decisions and subsequent activities are the most visible to the observer in terms of everyday accessibility in the field, and therefore are the ones easiest to gather data on. It would be easy to decide that the irrigation system in San Juan is decentralized, if only these activities were taken into account.

It is when the infrequent but crucial decisions in terms of long-range existence of the system arise that centralization becomes most manifest, for it is on these occasions, at least for the contemporary period, that the state becomes most plainly visible. It is in this context that the social embeddedness of irrigation system roles (the social organization of the irrigation system) is most crucial. Another significant point we are making in this paper is that in this case study the conflict resolution (and defense) functions are of major importance to the personnel in the system in terms of maintaining status inequalities, but also in terms of keeping the system functioning smoothly, and these functions are the most centralized. This centralization solves the problem of articulating local and national social systems through the double membership of the local elite in the local and national role systems. But these elite are not merely passive in their relationship to either system. Rather, they generate power both locally and nationally, and use the power generated in each domain to strengthen their position in the other.[4] Moreover, decisions based on perception of economic or political needs at the local level may contradict decisions based on perceptions of economic or political needs at the national level. This situation was clearly present in our case study, specifically in the context of the building of a new dam by the Papaloapan Commission. Thus, it is also of importance to show the role of local and national elites in mediating, fostering, or slowing down processes based on decisions made at different levels of the system.

## THE PHYSICAL SYSTEM

In the literature on societies which practice irrigation, one frequently finds statements which refer to "the irrigation system." What some authors mean by this term is the entirety of phenomena concerned with irrigation; some mean the society itself which practices irrigation; while others, such as Fernea (1970) usually mean what we are here calling the *physical system.* This ambiguity in a key concept has led to some confusion; in our view, it is important to distinguish the physical system from the social system. By the *physical system* we mean the relevant physical environment (e.g., amount of water available) plus the artifacts in and on the ground (dams, canals, sluices, etc.). By *social system* we mean the social organization connected with the control of the physical system(s). The size and other attributes of these two systems logically can vary independently of one another. The physical system can be as small as a single Ceylonese tank, or as big as the systems in the Mesopotamian valley, where the canals are often river size, covering scores of kilometers. Three logical possibilities exist concerning the relation of size of physical and social systems. First, the social system may cover all, and only, a single physical system. It is conceivable that a single physical system could support two or more separate and distinct social systems. (On a logical basis, it appears to be unlikely that this condition would be stable or enduring. Competition and conflict would probably be too extreme to permit two alien societies to share a physical system. However, two sub-systems [i.e., two communities of the same politically defined society] often share a physical system.) The third possibility is a single social system which deals with several different physical systems. This case exists where the politically defined society is a state, and the physical irrigation systems are small: e.g., systems based on small streams, wells, qanats and/or where tanks are the major form of bulking of water.

In this latter case, anthropologists have tended to focus upon the scale of the physical system to the detriment of a full understanding of the total scale of the social system and the relationship between physical and social systems. For example, Price's criticism (1971) of Robert McC. Adams (1966) can be traced in part to the fact that "the size of the irrigation system" may mean the size of the canal network (the physical system) or the scale of the centralized polity. This ambiguity results from the term "irrigation system." If the term refers only to the size of the physical system, then clearly Pul Eliya, for example, does not conform to a limiting condition in Wittfogel's model, the large irrigation system, but is, as Millon (1962) indicates, a small-scale system. But if "irrigation system" refers to the social system (i.e., the state of Ceylon, of which Pul Eliya is only a small cog), it is large in scale. A state may, and frequently does, contain a large social system concerned with irrigation, which consists of direct and indirect control of many small physical systems. Failure to attend to this distinction is a source of confusion in attempts to assess the relationship between irrigation and political centralization by the comparative method (cf. E. Hunt 1972).

### The Landscape

The Cuicatec region is one of differentiated ecological characteristics. From the canyons of the Rio Santo Domingo and Rio Grande (La Cañada de Cuicatlán or Tomellín) at 500 to 600 meters above sea level, the terrain slopes up to the uninhabited Llano Español plateau at an elevation of about 3,200 meters.

Basically, there are two ecologically different agricultural zones, the semi-arid Cañada del Rio Grande and the high temperate mountains to the east. The canyon, or Cañada of the Rio Grande, supports a xerophitic vegetation consisting of such plants as mesquite, thorny bushes (*cardoneras*), *cuajilote, palo mantecoso,* hard grasses, and *organo* cactus. It is characterized by low rainfall, high evaporation rates, and a large permanent river, the Rio Grande. The affluents of the Rio Grande are shallow, narrow mountain rivers with steep gradients which have deposited rich soil at their confluence with the main river.

These relatively flat alluvial flood plains are intensively cultivated with irrigation. Their fields produce 800 to 1,500 kilograms of maize per hectare per planting. Plums (*obos*) are raised, as well as chicozapote, mango, chile, tomatoes, some rice, and fodder. Sugar cane is grown commercially and processed locally. Habitation in the Cañada is found only in association with these alluvial fans, in locations where the canyon opens out into a relatively wide valley floor. Elsewhere the terrain is rough, non-irrigable, and uninhabitable. Of the towns and villages located in these fertile localities, all but one, Valerior Trujano, had pre-Hispanic antecedents.

Above the Cañada region, the lower ranges of the hilly flanks are unusable for agriculture. This area contains small animals and edible varieties of wild plants. Various wild seedpods and cactus fruits are still gathered here by the very poor. In prehistoric days this area was inhabited by hunters and gatherers living in rock shelters (E. Hunt 1972). The thin vegetation is a mixture of the forest and plants found in the richer areas above and below.

The mountain towns at altitudes between 1,500 and 2,000 meters are located on small, flat areas or plateaus, or on spurs alongside the streams. Here is found the bulk of the highland population; lands above and around the towns are used for seasonal rainfall crops. The land is cultivated by an outfield-infield system, which produces maize at the rate of 400 to 600 kilograms per hectare. The size of the areas irrigated is very small; sometimes only two or three parcels of land support an equal number of households. Irrigation is thus a very minor component of the agricultural highland complex, playing a supplementary role in terms of yields, unlike the canyon, whose role is basic for survival of the agricultural system.

The upper slopes are covered with a sparse primary forest of deciduous trees and secondary grasses. Above 2,000 meters are some high altitude agave plantations. Otherwise, a mixed forest provides firewood and lumber for house construction. Several species of pine are the source of cellulose used by the Tuxtepec Paper Co. The country is cold, sometimes touched with frost and snow.

In the remainder of this paper we shall concentrate primarily on the Cañada itself, at first summarizing the significant ecological contrasts. The Cañada climate is semi-desertic; the water balance aspects create extreme dryness from the point of view of agriculture. As the tables show, the Cañada is an arid system where there is a high net water loss to the atmosphere during every month of the year (Tables 12.1, 12.2, and 12.3). For six months of the year there is literally no rain at all, and the maximum number of months of drought (*meses con sequias* ≥ 28 *dias*) in a year is ten and eleven respectively for each of the two reporting stations. For six months of the year,

**TABLE 12.1* Relationship of Evapotranspiration to Mean Precipitation, in mm., for San Juan.**

| | Jan | Feb | Mar | Apr | May | June | July | Aug | Sept | Oct | Nov | Dec |
|---|---|---|---|---|---|---|---|---|---|---|---|---|
| Evaptr. | 80 | 85 | 145 | 160 | 180 | 175 | 160 | 160 | 130 | 100 | 80 | 65 |
| Precip. | 0 | 0 | 5 | 5 | 35 | 115 | 120 | 60 | 100 | 55 | 0 | 0 |
| P–E | -80 | -85 | -140 | -155 | -145 | -60 | -40 | -100 | -30 | -45 | -80 | -65 |

**TABLE 12.2* Useful Rain (Rainfall ≥ 24mm with 80% Probability) per Month, Quiotepec Reporting Station.**

| Jan. | Feb | Mar | Apr | May | June | July | Aug | Sept | Oct | Nov | Dec |
|---|---|---|---|---|---|---|---|---|---|---|---|
| 0 | 0 | Trace | Trace | Trace | 30 | 35 | 20 | 35 | 20 | 0 | 0 |

*Data summarized from Comisión del Papaloapan, 1956.

therefore, there can be no agriculture which is dependent upon an atmospheric supply of water. Furthermore, as the precipitation minima show, there are often whole years in which no agriculture based on atmospheric supply of moisture is possible.

The total supply of atmospheric water to the Cañada is in general quite low – the lowest of the entire Papaloapan Basin. Of even more importance, however, is the measure of usable rain. In the Cañada, there are in effect seven months with no usable rain, and of the remaining five, only three, on the average, have more than 24 mm. It should be stressed that these figures are means, and that in the case of minimal total rainfall, no usable rain over 24 mm will fall, if indeed any falls at all. Therefore, agriculture without irrigation in the Cañada has such a high risk potential that it can be said to be culturally non-viable. The people in the canyon say that agriculture without irrigation is impossible. In San Juan one man tried to grow plum trees above the highest canal, watering them by bucket, and everybody thought he was crazy. He was an outsider from the valley of Oaxaca where well irrigation is traditional, and when his venture failed, he left town. In fact, all agriculture since the sixteenth century has been based on canal irrigation.

A comparison of the Cañada with the slope and peak areas of the mountains to the east shows dramatic differences. In general, the higher slopes get twice as much rain, three times as much useful rain, nearly twice as many months of useful rain, only half as much evaporation, and not quite half as many drought months in a year. The highlands are cooler, cloudier, and more moist. These highlands support a sizable population based largely, though not entirely, on what is called temporal agriculture, that is, agriculture based on seasonal rainfall.

While the atmosphere rarely bestows significant moisture, the Cañada contains a large permanent stream, the Grande River, which is one of the sources of the Papaloapan. This stream runs, for most of its length, through a narrow canyon. Smaller streams and brooks flow down the mountains. Although some of these streams are permanent, they vary greatly in flow between the dry and the rainy season. The gradients of these streams, especially the small ones, are quite steep. The principal one for this paper, the Rio Chiquito, which flows through the town of San

**TABLE 12.3* Comparison of Cañada, Slope and Mountain, on Various Climate Variables.**

| | Cañada | Low Slope | Mountain Towns | Plateau |
|---|---|---|---|---|
| 1. Temperature – Jan mean | 22°C | 17° | 13.9° | 13° |
| 2. Temperature – July mean | 27° | 20° | 17° | 17° |
| 3. Insolation – January | >70% | ? | ? | 40-60% |
| 4. Insolation – July | 40-50% | ? | ? | 40-50% |
| 5. Evaporation – annual | 2750mm | 2500-2000mm | | < 2000mm |
| 6. # Months with ≥28 days of no rain | 11,*10*† | 8 | 7 | ? |
| 7. Mean annual precipitation | 570mm | 800mm | 1000mm | 1500mm |
| 8. Total annual useful rain ‡ | 135,*146*mm† | 200mm | 500mm | 500-1000 |
| 9. # months with useful rain ≥24mm. | 3,*4*† | 5 | | 6 |

*Data summarized from Comision del Papaloapan, 1956.

† In the Cañada, there are two reporting stations, thus the double response. San Juan is *underlined.* Gaps in the chart are due to lack of certain figures for certain areas of the region.

‡ Useful rain is defined as 75% of precipitation over 5mm, for each time it rains. 24mm were chosen as base line, for the Commission used this figure of necessary water for corn, derived from figures for the State of California.

Juan, drops 2400 meters in 23 km. These streams are young geologically, and have narrow rocky beds for most of their length. The streams frequently cross aquifers located along the whole length of the bed, whose seepage accounts for a large percentage of the total flow of the stream.*

The Cañada valley system soils and terrain are for the most part unusable for irrigation. The loci of cultivation and human settlement in the Cañada are the alluvial fans of the mountain streams, which represent a small percentage of the valley floor, and are separated from one another by stretches of narrow, uninhabitable canyon. Within the whole length of the Cañada, including the Vueltas River, there is a string of eleven settlements, beginning with Atlatlauca in the south and ending with Quiotepec in the north. The major settlement is San Juan, which is discussed in detail in this paper.

The agricultural lands in the Cañada are irrigated by small, closed canal systems which service every settlement. For a system of irrigation of this type, three different characteristics should be considered: first, how the water is extracted from its natural source; second, how it is transported to the fields; and third, how the water is made available to the plants which utilize it. In the San Juan area, the conditions are among the simplest. The major sources of supply of water are the fast running, steep gradient streams; a secondary source is the Grande River itself. Diversion occurs far above the fields; water is transported by gravity flow down small, unlined ditches. This system simply continues into the fields in small subsidiary ditches; there is no bulk storage of water for irrigation.

There are some significant human consequences of these environmental conditions. The diversion dams have to be strong enough to withstand, for a reasonable amount of time (several months) a fairly rapid stream flow. The distributory canals are by and large short (the longest is 10 km., most are less than 5 km.), and collect relatively little silt when compared with nearby systems such as the Tehuacan Valley, or with other systems such as those in Iraq, in which silting is a major problem. Flow in the canals is good, as it is into the fields, if the canals are cleaned regularly. There is therefore relatively little demand for an input of complex technical knowledge and esoteric skills; however, the labor input of such a system is not low.

Since each settlement has, typically, sole and undisputed access to its stream and alluvial soils, each community tends to depend on a physical system independent of other communities. That is, there is a very strong tendency for physical system to be community specific, which means that the take-off, or diversion dam, and the transport system are entirely within community boundaries, and further that all the water in the system is utilized by the community. Given the abundance of aquifers, there is little if any opportunity to completely deprive a downstream community of water. It is also possible for some communities to have two or more independent physical systems operating simultaneously, due to a plethora of very small sources.

However, there is a major problem created by low water supplies, given the demands of the population. During the hot dry season from December to the beginning of April, water for irrigation is scarce.[5] Several factors besides lack of rain increase scarcity. Sizable amounts of water are lost by evaporation and absorption. Part of the stream beds are porous sand and loose stone, and in some cases 40 percent of the running water is lost. As might be expected, water in this area is and was extremely valuable; the need to control water supplies is a powerful influence on the social structure as we shall see below (E. Hunt 1972).

In the Municipio of San Juan, the major area of agricultural fields in the Cañada, the physical irrigation systems function without any impressive or major permanent physical superstructure attached to them. Two types of permanent water works exist which are of marginal utility. Within the town of San Juan, there is a stone-faced (and partially covered) network of street ditches, which is utilized to provide household water; part of this water is used to irrigate fruit trees in house yards. This arrangement exists only within the few central blocks of the contemporary town, is kept in a high state of disrepair, and is of colonial origin. Each household keeps up its own branch of the ditch as it sees fit. Outside this miniscule system (servicing approximately ten blocks), the colonial government built several small Spanish-type aqueducts in the typical arch design to service a few towns in the Canyon area (Martinez Gracida 1883). These aqueducts have either been left to collapse or are used in a near total state of decay with loss of water and inefficiency of distribution.

The bulk of the agricultural lands today are irrigated by extremely simple techniques. Several major feeder canals (*apantles,* from the Nahuatl word *apantli*) serve the fields and orchards which surround the town of San Juan itself. Minor irrigation ditches spread over the irrigated land in the municipio like a spider web. These ditches are open and shallow, with

*Papaloapan engineer 1964: personal communication.

mudpacked walls. The deepest is one meter, but most of them are between 30 and 40 cms. Where topographically necessary, a simple rock or brick buttress against a cliff face supports them; here and there a metal pipe or a hollow log (called a *canoa*) takes the canal over a break in the terrain. Gates and sluices are temporarily built of piled-up, uncut stones, mud, and branches. *Tomas de Agua,* the main diversion openings or "dams" from the stream, are built of mud and branches. Larger ones are built of chains of small pyramids made of piled-up branches stuck in a semi-vertical position. These "dams" are locally called *chalchihuites,* a Nahuatl word which means "green stones," or "necklace of precious green stones," and is a pre-Hispanic poetic symbol associated with surface water and its Goddess Chalchiuhuitlicue, companion of Tlaloc, the rain god (Florentine Codex 1953:Vol. I:6-7; Caso 1953; Piña Chan 1960:80-81, 84, 115). Chalchihuites are easily washed away by freshets or strong currents, particularly in the rainy season, so that constant servicing is necessary. Cleaning the canals and opening or barricading sections of ditches and the primitively built chalchihuites are simple but tiresome and frequent tasks. The number of man-hours involved in those services varies from one case to another, depending on the size of the canals and the form of canal "tenure."

To maintain minimum standards of efficiency, decisions on when and how well to repair the system, especially in the rainy season, require skilled decision making. Waiting too long to repair a dam may result in discovering that it has been totally destroyed by the strong current. If repairs are made too soon, the labor costs are often higher then necessary, thus deflecting scarce resources from more productive pursuits.

In San Juan, two town canals using water from a small permanent stream, the Chiquito River, cross the residential area above the central part of the town. One services the house orchards and land around and below town; the other services lands north of the town, in a section called "The Grassland," where most small land owners have their parcels. The southern canal is again divided into two major branches; one of these crosses the town southward, the other in a western direction.

Within the municipal boundaries there are two major areas of agricultural fields which are not irrigated by publicly controlled canals. The best land of San Juan, adjacent to the Grande River, is located outside the range of irrigation of the town-controlled ditches. This area, called "Spanishland," is an extremely fertile flat shelf which edges the alluvial fan of the Chiquito River, parallel to the Canyon lower spurs below the railroad line. One major feeder canal drawing water from the Grande River services all the Spanishland to the north through a complex network of ditches arranged in a perpendicular and parallel rectangular pattern. This pattern is quite unlike the town canals which, being located in poorer, rougher terrain, follow natural features of the landscape, fanning outward from the major canals in a branching pattern. The southern section of the Rio Grande shelf is irrigated by another canal (San Pedrito) which originates in a small stream "owned" by an Indian barrio which is a municipal dependency of San Juan. This canal bed was built in a previous generation and is kept in working order by one of the elite families of San Juan, primarily to service their own lands.

Another main canal drawing water from the Grande services the small ejido-controlled lands further north of the town.

## THE SOCIAL ORGANIZATION OF IRRIGATION

### The Social Structure of San Juan

In this section, we will be primarily discussing the relevant aspects of the social structure of San Juan as a semi-autonomous community, but we will have occasion to refer to the national-level system, of which a brief description follows. Mexico as a nation is a federal entity, composed of states (*estados*). These states are sometimes divided into districts, and always subdivided into semi-autonomous territories called *municipios.* The political institutions of Mexico are strong and highly centralized in a single party system and a unified federation controlled from the capital of the nation (cf. Gonzalez Casanova 1965; V. Padgett 1966). Of primary importance in this context is the government itself, with its various bureaucracies which reach down to the local district and municipal levels with varying degrees of effectiveness. In addition to the courts and tax collecting systems, which may affect water usage, there are also agencies directly concerned with the administration and management of water resources. One is an agency of the Federal Ministry of Agriculture and Livestock, the other a Ministry in its own right, called Recursos Hidráulicos (Hydraulic Resources). The watershed of which San Juan is a part has been centrally organized and developed since 1946 by a National Federal Commission (Comisión del Papaloapan), which has had responsibility for planning, construction, and management of a large series of water works including the major Aleman dam and, at the local level, small diversion dams, other irrigation works, and tanks for potable water. This commission has its own federal budget, and works with state and local government in coordination. It has sponsored research (Attolini 1949, 1950), published various kinds of data relevant to irrigation works (e.g., Comisión del Papaloapan 1956), and even had a history written on it (Poleman 1964).

The other central political institution which is of major importance for this paper is the political party in power, the Partido Revolucionario Institutional (the PRI). This party is a more or less unified, enormous bureaucracy, which parallels the government but is not identical to it (O. Paz 1972).

The major local unit of government and political organization in San Juan is the Municipio which is incorporated with all other Cuicatec settlements into judicial and electoral district. The Federal Constitution of 1917 and the state law codes guarantee its local sovereignty in internal affairs of the community such as elections of officers. Local political agencies and institutions are therefore legally under the control of local citizens, who make many everyday decisions for the locality. San Juan also has a Court of First Instance, which gives it a local judge who also heads the judicial district. There is, therefore, something of a disjunction between the central federal and state governments on the one hand and the government of the municipio on the other. At the same time, however, the PRI has a local branch committee at the municipio level. Its leaders, a fact which is repeated all over Mexico, are the locally powerful people, powerful both in wealth and access to outside sources of support in politics. Part of the political redistribution system in Mexico operates by local committeemen of the PRI funneling substantial campaign contributions and personal support of their clients up the PRI channels. In return, members of the local elite receive support for their candidates in nominations for elected and appointed positions in the government, and other kinds of special considerations. The PRI itself, then, is the one way in which local social systems are articulated with the national ones at the politcial level, but outside the government apparatus itself.

The national landscape, as we will see below, is relevant in a discussion of the social effects of irrigation. But, before we turn to a discussion of the specific role systems and other social phenomena directly connected with irrigation, we must present some important features of the social structure of San Juan. Primarily, the following section focuses on the differences between the social classes of San Juan, differences which are directly related to the uneven access to water between the town's social groups.

The town of San Juan is the capital of a municipio, and the capital of an *ex-Distrito.* During the days of the Porfiriato, the state of Oaxaca was divided into a score of districts, each of which was administered by a political chief (*jefe político*) appointed by the central government. With the reforms of the Revolution, the municipios were granted a degree of political autonomy, including the absence of centrally appointed district administrators. Today the districts do not officially exist as a territorial units, but a unit called the *ex-Distrito,* which functions for judicial purposes, census taking, taxation, and so on, is still extant. Its functions are much attenuated from the past, but it has continued to be a socially significant unit. The major importance of the district, from our point of view here, is that the capital of the district contains many federal offices, such as tax, postal service, and court. These offices provide a listening post for local elite, opportunities for patronage appointments, and easy access to this level of the federal and state bureaucracy.

As we mentioned, San Juan is also the administrative center (*cabecera*) of its own municipio. The state of Oaxaca has by far the largest number of municipios in all Mexico. As a consequence they are generally much smaller in area and population than in the rest of Mexico, especially in the Cuicatec Indian area, where most municipios have fewer than 5,000 people, and are rather heavily nucleated and small in area. The Municipio of San Juan is unusual in that it contains most of the population of the Canyon in several settlements spread out over a distance of more than 30 km. Moreover, the Canyon municipios, unlike the highlands, have been populated by Mestizos, (i.e., Spanish-speaking nationals) since at least the late seventeenth century. The civil-religious hierarchy which controls much of the political life in homogeneous Indian municipios does not exist in the canyon towns. This circumstance has direct effects on the social organization of irrigation (cf. S. Lees 1970).

As the *cabecera* of the municipio, San Juan contains considerably more division of labor at the official level than do other, smaller dependent settlements in the municipio. The town is governed by elected municipal officers who remain in service for three years. In addition to being a center of intensive irrigated agriculture, San Juan is also a commercial, banking, transportation, and service center for the district. One of the two priests in the district is located here, as well as the only doctors, a dentist, the only medical facility (run by Social Security, a federal agency), and the major wholesale establishments of the district. Most San Juaneros are involved in agricultural tasks primarily for the production of irrigated cash crops for the national market and in business activities involving import and export trade with the Indian hinterland and the larger urban centers of Mexico.

San Juan contains three distinct social classes, which have different positions vis-à-vis control of land and water. There is a small elite (*la gente de categoria, la clase alta*) comprising approximately 10 percent of a population of 2,500 inhabitants, a small middle class (*la clase media*) of approximately the same size, and a majority of the lower class (*los peones, los de menos*).

The elite in town are the major owners of land and commercial establishments. They control the bulk of the best irrigated land in which they plant sugar, rice, and fruit trees; they own all of the major stores and market stalls. The elite are organized in terms of corporate kindred, which are homogenous with respect to class and political faction. These kindred are inter-married across town boundaries; such kinship networks form the basis of cooperating economic groups and political support, and spread, territorially, into other irrigated towns in the district and wider region, and into major metropolitan centers (cf. E. Hunt 1969; E. and R. Hunt 1969; R. Hunt 1965, 1971, for more extensive discussion of these issues).

The middle class derives income primarily from (white collar) service jobs, and to a degree from retail commercial activities and minor cash cropping primarily of fruits and sugar. The local doctors, druggist, and dentist are a part of this class, as are the full-time bureaucrats. Relatively well-paid full-time clerks and assistants in the stores and the owners of small retail stores are also in this category. Ownership of irrigated property is present, but is not of major importance to the income of the family compared with the upper class. The members of this class disdain manual work, and tend to devalue agricultural pursuits, unlike the elite, which is intimately involved in agricultural and water management.

The lower class includes a roster of low income San Juaneros, from the owners of small parcels of irrigated land sufficient to produce cash to provision their family to completely propertyless individuals who must work in agricultural tasks for wages. Individuals in this class own some minor cash-producing property, such as a very small field, or an inherited share in the profits of the harvest of a fruit tree. It is rarely the case, however, that low-class families can live solely on the proceeds of their property. Their major source of income is peon work, manual labor connected either with the soil (work in fields, or on the irrigation works), with construction, or as human burden carriers (movement of goods from a store to the house of customers, moving items in a warehouse, helping to load or unload a train boxcar, etc.).

Many other social features correlate with these three Weberian classes. Education beyond the sixth grade level is limited to the two upper classes. Only the elite and the top of the middle class have access to university levels of education. Life style correlates directly with income and education, as does the kind of knowledge and experience the person has of the outside world. Critically, the members of the elite and middle class are the only San Juaneros who, on a regular basis, cope with the metropolitan world, including the higher levels of the government machinery.

San Juan also contains two political factions. These factions are essentially struggling for control of the basic wealth of the community, which is composed about equally of the irrigated fields and their products, and of extensive commerce with the Indian hinterland. The division between factions is very strong among the elite, less strong among the middle class, and loose enough to be an opportunity for manipulation of clientship "privileges" (including access to water) among the lower class. There are virtually no occasions on which the elite cooperate across factional lines. The two exceptions involve control of irrigation and taxes as we shall see below.

The stranger is struck by the number of government jobs which exist in San Juan. Many persons in town are full-time government employees, from the *mozo* (field hand) who sweeps out the jail to the district judge. Many others, it turns out, work in an official capacity on a part-time basis. The municipal president is usually a member of the upper or middle class, and such posts as jefe politico or now congressman have seldom left the hands of the upper class. Of even more importance, however, is the fact that the vast majority of persons occupying government offices do so to some extent at the pleasure of the local elite, either as a consequence of decisions made within the local PRI committee, or as a consequence of powerful patron-client and *compadrazgo* relationships that local elite have outside the town with powerful persons in central bureaucracies.

We must turn now to some formal aspects of water control, and its connection with the class structure. An important feature of states, frequently ignored by anthropologists engaged in community studies, is the relation between local norms and regulations and the codification of rules of access and alienation of property and resources by a literate central bureaucracy. This relation is eminently important in this case study.

In Mexico, water is owned by the nation and in theory is available to all potential users in an egalitarian fashion, which involves taking regular turns at being served and paying identical rates of taxation. But the ideal picture of an egalitarian system of water distribution, maintained by an egalitarian system of duties with respect to the physical maintenance of canal networks, is complicated by a complex system of water or canal and land tenure and different actual water demands of the crops planted by the different canal users. Land in San Juan is owned privately by individuals, by family trusts, by private corporations (a mill), cooperatively by the local *Ejido* members, and by the town corporation to grant in usufruct. Some land can be rented, leased by long term contracts, or mortgaged. An additional complication is that permanent standing crops such as fruit trees can be sold, rented, or leased independent of the land on which they stand (the origin of this practice is pre-Hispanic [Millon 1955]).

To complicate this normative picture, there are irregular and illegal aspects of the social organization of the irrigation system which are nevertheless omnipresent. The way in which the physical systems are actually administered is an extraordinary example of the divergence between normative structure and actual practice frequently observed in other aspects of Meztizo Mexican village life. It is important that this point be stressed, for accurate field data are crucial and easy to miss on account of this divergence.

Whenever asked about conflict or bribing in regard to water, which is endemic in San Juan, most informants act, if not well acquainted with the anthropologists, as if blind and deaf. Admitting knowledge of the illegalities occurring in the system is tantamount to admitting that one has engaged in them oneself. Thus, getting lists of irrigators, of true water schedules, of payments, or accounts of specific resolutions of conflict, is difficult in the extreme at the beginning of field work. It was only by living in the town for a sufficiently long time and acquiring access to intimate, private information, that we discovered how the social organization of irrigation in fact works, rather than purely how it is supposed to work officially. The wife of one of our informants put it quite graphically when she said that her ". . . husband will not talk about water business, not even to himself when he is asleep." This point is an important one, because a picture of water control in Mexico based on interviews during brief stays in communities (for example, using information obtained in surveys) will give only a formal, ideal picture of how water-control systems are expected to work if legal, normative arrangements were followed. But the actual picture of water control is very different from normative expectations.

At present there are several different social agencies which control the physical irrigation systems. There are "private" canals, a public, town-controlled canal network, an Ejido canal system, and a drinking water system.

## The Canals of the Town

The two town-controlled canals service the majority of users of irrigation water.[6] Theoretically every town resident who desires can have access to town water for irrigation purposes. In fact, only those who have lands located in the path of the canals and cannot afford any other water source pay the town for its water. The town-controlled canals and water are administered by a specialized agency whose officers have no other authority; that is, their roles are exclusively defined in terms of water control and formally isolated from other political roles in the community. This agency is called the Water Commission (La Junta de Aguas). Officially, it is a branch of the federal government which owns all national water sources.

The present Water Commission is administered by a governing body whose officers are elected annually. The officers receive water payments from users. The income of the Water Commission (approximately $500 monthly) is used to defray the costs of the town-controlled sections of the irrigation system, including the salaries of the Commission employees. The Water Commission also decides the price of water according to the degree of scarcity during different seasons, adjudicates conflict cases between water users within the village, keeps schedules and general bookkeeping of irrigation turns, and decides when to ration water by the hour in times of scarcity.

Nobody in town wants the administrative jobs in the water commission except the *Fontanero* position (see below) because they give no advantage in water control for the men in charge. Officers are constantly suffering from pressure by different water users to favor their lands. The poor serving in office find that they cannot resist the pressure of the larger landlords (who are usually their patrons and creditors). The rich find no advantage in service in the committee themselves because they can delegate their power to their clients and/or irrigate from other sources. Thus, no one wants to monopolize these positions. One informant said that he found the service unprofitable because it involved innumerable headaches: three or four hours of daily evening meetings, hearing constant complaints, and "not even the possibility of getting more water for oneself."

The commission officers thus, basically, hold temporary power by being official representatives of the community as a corporation vis-à-vis its members, and because they do not interfere with the access to water of the politically powerful and land-rich elite. Basically, the Water Commission's authority is derived from the strength of the community as a whole, indirectly backed by the centralized federal state. The officers of the commission hold this exclusively by maintaining consensus among the small landowners who use town water, and by not antagonizing the large landowners.

The commission's governing body has a president, a secretary, a treasurer, five committeemen (*Vocales*) and two water policemen or distributors (*Fontaneros*). The president is responsible for reports to the membership, and for decisions on cooperative labor for canal cleaning and repair; he also acts as liaison with the federal offices of the commission. The secretary acts as scribe and record-keeper, copying irrigation schedules, bills, and records of decisions made by the membership in open meetings. The treasurer keeps the money and the accounts of cash transactions. The *Vocales* act as representatives of the body of town water-users during regular meetings.

The water policemen are officially in charge of the two town canals. They open and close sluices. They make everyday decisions about where and how long to irrigate specific parcels by determining how dry they are, and whose turn is coming up. They police the sections of land being irrigated, guarding the canal sluices armed with guns. They are expected to arrest those who they find stealing water, and carry them to the local jail. Each one of the water policemen is supposedly in charge of one of the canal branches, taking turns at work when water is being diverted to their ditch. In fact, however, they work together to protect each other from attacks by irate water users who find themselves in disagreement with their water allocation policies.

The work of the *Fontanero* officers is thus the most dangerous, and also the only profitable kind. If skillful, they can make a substantial income from bribes, mostly during the dry season, when everyone tries to gain an advantage over the neighbor in irrigation turns. Moreover, they are major adjudicators on a day-to-day basis on conflict between parcel owners, and are thus called water judges (*jueces de aguas*). Hence, although their regular monthly salary is only 40 pesos, it is said that they in fact get enough money from illegal payments on water to live on their income.

Water in the town canals is legally cheap and because of this, salaries for the water officers are low. Water from the town canals is illegally expensive, and because of their low salaries, water policemen are expected to overcharge, particularly during the periods of water shortage. Even after payment, a man has to wait near the ditch when his turn comes up, to make sure he will not be by-passed. Many do so armed with guns or machetes for insurance. Official payments for water are made as fields are irrigated, according to the prearranged costs on a *pro rata* basis. Water can be paid by the hour. The most common agreement, however, is payment by amount of seed measured in *arrobas* or *maquilas*, a standard calculation being that an *arroba* is equivalent to 1/4 hectare. The price per *arroba* is modified by the crop planted, which San Juaneros feel is just, since some crops require more water than others. Corn, the cheapest crop, runs about 1.50 to 3 pesos per *arroba*.[7] Tomatoes, which require about three times more water and are a popular crop, cost at least 4.50 pesos per *arroba*. A cultivator calculates how many *arrobas* he plans to plant, and makes an agreement with the Water Commission officers beforehand, ascertaining that he will be able to get sufficient water for his crop, and agreeing on a minimum annual payment for it. Fruit trees, however, are irrigated about every twenty days by paying for a certain number of hours per watering. These have been one of the most important town export crops from pre-Hispanic times. These payments are permanent costs for orchard owners, since mangoes and chicozapotes produce annual fruit for longer than a human lifetime. Everyone in San Juan, irrespective of class, tries to own fruit trees.

Receiving water also implies an agreement to contribute laborers or one's own effort to canal maintenance and cleaning, but as we indicated, these payments are separate from payments made for specific waterings. Prices of labor also vary with the crop planted. Sugar cane requires twice as many man-hours of labor as corn (3 to 6 laborers per hectare). Thus a man drawing water from a public canal is expected to provide maintenance service proportional to his consumption of water. The actual work of cleaning, however, is done by the lower class. Large wealthy land-owners, when drawing water from public canals, hire men to perform their share of the cleaning. Small land-owners and renters do the work themselves, sometimes with the help of kinsmen. During the planting and growing season, crews of *mozos* (field hands) are regularly at work under the supervision of canal owners, foremen employed by them, or officers of the town Water Commission, who are sometimes aided by the municipal authorities. Such work crews are primarily concerned with the maintenance of the system, that is, cleaning and repair of canals and sluices.

We have not attempted to calculate the total number of man-hours necessary to maintain the whole of the canal network of the town because, early in the fieldwork, which was focused on other problems of research, it became obvious that we would not have the time to collect all the necessary data. However, various figures are available. Canal cleaning, digging, and repair, where controlled by the local town water authorities (i.e., the Town Water Commission), takes up the larger part of the total communal town labor. One single cleaning of the Grassland branch canal required two full days of work for 115 men, which at the local average of 6.5 hours per day amounts to about 746 man-hours. For maintenance of the few large, privately controlled canals, one or two *mozos* are permanently hired for continuous work. Extra crews of five to ten men are routinely hired for major cleaning and repair; they may also be hired for various periods of time to take care of emergencies. Several permanent crews are in charge of opening and closing sluice gates which divert water to different fields. These crews are constantly present when water is being allocated to prevent farmers from stealing it. In seasons where water is in short supply, stealing water is very commonly engaged in by owners of neighboring parcels of irrigated land, especially at night when visibility is poor. Therefore, most townsmen prefer to irrigate during the daytime, even though the volume of water per hour is greater at night due to a much lower evaporation rate.[8] Individuals' claims to rights of irrigation water are validated by residence and town membership, and by paying the dues or taxes for services of the Water Commission. But payments to the Water Commission are not made unless one cultivates land which is irrigated by town-controlled canals. Thus, although in theory the town as a whole controls the Water Commission, in fact only users of town canal water actually have any indirect control over distribution and allocation of this water. Users, however include both landowners and renters or sharecroppers, the last two acquiring water rights indirectly from the landowner's potential rights, and directly by payment to the Water Commission.

Water Commission members have to add to regular payments and taxation a certain amount of labor on their own to keep clean the subsidiary ditches which reach their parcels; the town does not assume this responsibility. Usually, water users controlling a smaller ditch in common clean sections in cooperation or by agreeing to provide a certain number of man-hours (themselves or peons) for upkeep of their ditch.

Although the Water Commission does not keep up subsidiary canals, it has the power to enforce upkeep by a ditch user. If a man refused to cooperate with the neighboring parcel owners and keep up with his share of labor, the other irrigators can complain to the Water Commission and insist that water be refused to him until he complies with his proportion of work.

All these agreements made in or through the control of the Water Commission do not represent the bulk of the irrigated land. Most of the land irrigated by the town canals is in household orchards inside the town. The major area of parcels of the Grassland has a total surface of less than 1/3 of the cultivated land of the town, although all but a handful of cultivators plant there. (There are over 200 water users in the commission.) Parcels are very small; except for the center of the area, most parcels cannot be irrigated in the dry season because of water scarcity. This problem is one of the major sources of internal strife over water, which is described in the section on conflict.

## The Elite Canals

The Spanishland and San Pedrito canals irrigate the bulk of the best lands of San Juan. Spanishland controls about one hundred and ninety hectares of the best alluvial land. The private canals control a minimum of sixty hectares of alluvium, but errors and omission in our data (as compared to the map of irrigated land) suggest that perhaps the real figure is above one hundred hectares for both cane and fruit trees. All of these lands are controlled by the major landowners in cooperation with the owner and administrator of the local sugar mill. The Spanishland canals and ditches are controlled by a separate Water Commission administered by the sugar mill for the large landowners who plant sugarcane they sell to the mill. Priority rights on buying the sugarcane harvest are given in exchange for credit on labor, equipment, and water. The mill administrator himself makes decisions on both repair work and water allocations. The second canal of San Pedrito is controlled by a family of descendants of the original builder. The water is paid for by this family in the form of local taxes to the Indian barrio officers.

Payments in the Spanishland private canals are measured in *tareas*: a person pays according to the number of *tareas* his land requires. A *tarea* is the number of hours which it takes to plant or weed a piece of land approximately 10 m. x 10 m., or the equivalent time in canal work (i.e., it is a measure of labor in time).

These major canals and their subsidiary ditches are "privately" owned and "privately" controlled, irrigate the best land, and are all held by a small handful of major elite landowners. Approximately 190 hectares of these lands are irrigated by the mill canal. Of these, 150 are owned by five extended families of major landowners, four of which belong to the political faction which at present controls town politics. There are, however, twenty different landowners in this section. The Indian Village canal, which irrigates over sixty hectares, is also controlled by one of these four families, but serves more than ten different water users. Cultivators who do not belong to the town Water Commision membership because their lands are marginally located with respect to the town canals do not get water and thus do not pay taxes to the town for water. Since they are small cultivators and cannot afford to maintain their own canals, they buy the water (an illegal practice) from the private canal owners. The owner pays for canal maintenance and policing, schedules and keeps books on the distribution himself, and hires salaried policemen to watch night water turns. Thus, officially, water is not sold (which is illegal since waters are federally owned). What is sold by the mill and private canal owners are services in canal repair, policing, construction, and bookkeeping, as well as credit. These activities are a direct source of political power. All water users in private canals are political and business clients of the men who control these canals. Usually these users contract to sell their crop to the mill or the private canal owner (under a system called *refaccionamiento*). If a particular user does not "pay his share" in bribing costs, or does not support a political candidate, or refuses to sell his harvest to the man who gives him credit and water, he can easily be punished by cutting off his water supply. A man joining the opposition is sure to find sooner or later that he has been discovered, which means the ruin of his crop or the sizable reduction of his basic income (cf. Simpson 1937:370-71 for a comparable case). In fact one water client who joined the opposition was discovered when we were in the field.

Water selling in itself is a quite profitable enterprise. Private canal owners at times prefer to sell water and leave part of their own land fallow if they do not have sufficient water for both their water clients and themselves. In the case of the mill, the administrator is definite about the necessity of the mill for unified water control. If the mill doesn't manage the canals, he claims, fights between landowners cut down the sugar harvest. It is thus to the advantage of the mill owners to ensure that there is a just distribution of water among the large sugarcane planters. Only by maintaining the support of the growers can the mill function at a profit. With the exception of the present administrator (who is a professional), all mill administrators since the middle of the nineteenth century have been senior men of the San Juan large landowning elite families. As one of my landless informants phrased it: "They keep the water and the jobs among themselves."

But water selling can only be made profitable for those with a secure source of capital because initial investments and canal upkeep can be quite expensive. A single day of repair of the diversion dam for a branch of a private canal cost its owner 500 pesos (40 dollars) just in labor costs. Another repair of a branch dam cost 350 pesos, including the labor of seven men for five days. The cost of operating one private ditch was calculated to be 8,000 pesos per year. Such manipulations of the allocation of water are possible because of extreme competition over a scarce, lifegiving resource.

The Papaloapan Commission (cf. Poleman 1964) has tried to solve the problem of water storage (which greatly affects production levels for the district) by proposing a diversion dam and a new canal from the Rio Grande which will increase the amount of water available for irrigation and increase the area of land which can be regularly irrigated by a maximum of two-thirds of the present area in Cuicatlan. This project has been looked upon with jaundiced eyes by the local upper class, which perceives that it will cut into an important source of income and power. At the time that we were in the field, the upper class (unified in this issue across otherwise irreconcilable political factions) was exercising a coordinated effort to interrupt the building of the dam, both openly in town meetings and sub rosa by secret lobbying and meetings.

### The Ejido Canals

A third separate canal irrigates Ejido lands further north. The Ejido irrigates from Rio Grande waters, and is controlled by a separate set of officers elected among the Ejido members. Officially there are twenty Ejidatarios; unofficially there are thirty-seven separate parcels, amounting to 113 hectares, or approximately 3 hectares per Ejidatario. This canal is administered in a third, and different, arrangement. The officers in charge of water allocations, conflict resolution, and ditch cleaning are not separated from other roles connected with the Ejido. Water-control functions are assigned to Ejido officers as part of their general functions. Ejidatarios are primarily members of the lowest class. Their economic links with the elite are not based on water – which, uniquely, is independent of the elite – but on a dependence upon the elite for credit and sale of their products. Ejidatarios are also the regular skilled laborers at the sugar mill, near which many have their homes. This small handful of members of the lower class is disproportionately powerful politically since they are doubly protected by the federal government, both as members of the mill union and as the only members of the lower class with independent (Ejido) access to water.

*The Town Tank.* The town also controls the potable water tank, which services the central part of town through modern metal piping. Again, this water service is not universally utilized. To receive water into the house, special taxes have to be paid initially to extend the pipes, and annually for water received. This service has been introduced by the federal government through the Papaloapan Commission. Since it only services the central, most urbanized part of the town, it is basically a service provided by the town for the richest households, all of which are located in this town section below the canals. The poor live in wattle and daub houses located on the dusty, hilly slopes of the town above the canals and without direct access to household water. This inequality of access to potable water had no social importance in terms of class conflict because, as everyone in town was aware, the tank water was not any better than ditch water: the builders had forgotten to install the appropriate filters. In fact, the tank, with its coat of thick green slime and mosquito colony, looked mighty unsalubrious.

Thus in San Juan several separate arrangements of water allocation, related to separate loci of power, exist simultaneously. One set of loci is the control of each single physical system, one is at the level of the minimum autonomous political unit (the village, the municipio), and a third is at the level of the total society. In terms of official arrangements, the system is unified under the Water Commission, representative of the town of San Juan as a whole, but not of the municipio which should be formally the unit (the municipio contains other settlements and physical systems). The Water Commission is responsible directly to the federal government. In terms of actual arrangements of water-controlling roles and agencies, three separate canal groups operate: the Water Commission, the Mill Commission with the private canal owners, and the Ejidatarios. Over these lords the town's municipal government, which can bring water cases to court in time of conflict, into his own municipal court, or to the district court also located in the village. Above the municipio is the Mexican federal government which, through two of its separate ministries and other government institutions (e.g., the Ejido Commission), can reverse local decisions, arbitrate water disagreements between villages or within village factions, and introduce changes by building major water works.

Hence, water control is located in a set of institutions which are territorially localized and hierarchically nested in a vertical ladder of increasing power. This power ladder is isomorphic with the political and economic power ladder of the region and of the national state, and thus is ultimately centralized in a normative pyramidal arrangement, but simultaneously decentralized and semi-autonomous at the local level in terms of control of each physical system.

## THE HISTORICAL EVIDENCE

The picture of water control in tribal societies, or of Post-English Colonial Iraq, or of modern rural sectors of complex states such as Japan, cannot be directly used to explain the original evolutionary conditions under which water control did or did not become a significant social factor. It is obvious, however, that we can make some inferences about the universal requirements of irrigation under specified conditions by utilizing cross-cultural comparative materials and diachronic information in historically well-documented cases. Such materials may enable us to interpret cases which appear to be affected by the same distribution of variables. It is in this context that we wish to present historical information from the Cuicatec. We believe that our analysis of this ethnographic example suggests hypotheses of value to a clarification of some issues connected with more general problems of the analysis of the impact of irrigation on society. The following comments, therefore, are only hypotheses; generalizations are not legitimized by the nature of the data available for comparative purposes in terms of the typology we set up at the beginning of this paper.

Evidence from ethnohistorical materials indicates that the Cuicatec states were already developed at the time of the fall of Tula (ca. 1200 A.D.). Our only evidence from pre-Hispanic canals, however, comes from the post-Classic period, when the town of Cuicatlan was already a major semi-urban center, with a probable population of 5,000 inhabitants (Hopkins 1970; E. Hunt 1972b).[9] Thus the Cuicatec evidence at present cannot add anything to the question of the origin of Canyon states in relation to irrigation. We have extensive evidence, however, on the role that the Cuicatec ruling elite had in the socio-political life of their communities during the sixteenth century and the period just before the Conquest. This evidence firmly supports the view that the provincial rural elites of Mesoamerica, such as the Cuicatec, were actively engaged in providing leadership for the management of the agricultural affairs of the polities which they ruled.

At the time of the Spanish Conquest, the Cuicatec were a loosely integrated ethnic and political unit, composed of several autonomous town and city-states of Cuicatec-speaking peoples, each ruled by a native elite. They were a small provincial society, part of the traditions of Mesoamerican civilization, but geographically isolated from expanding culture centers. This small society was a wealthy and flourishing center of provincial life. Its economic cornerstone was intensive, irrigated agriculture, adapted to this harsh, inhospitable ecological niche.

The two different habitation zones which exist today existed in the sixteenth century. Territories of the communities in each, with few marginal exceptions, have been continuously occupied until the present, both in the semi-desertic Grande Canyon and the piedmont. Hardly anything that the Mesoamerican native planted can grow without irrigation in the Canyon. But with irrigation, the Canyon is much more productive than the mountain hinterlands. In pre-Hispanic time, population pressure led to warfare and attempted invasion of the Cuicatec-improved irrigated fields. In the Canyon, the several pre-Hispanic city-states grew in the center of their permanent orchards and agricultural fields, and their wealth attracted unsuccessful conquerors until the last Aztec king took the major city-states under the empire for tax tribute.

The Cuicatec masses were primarily farmers, cultivating most of the known Mesoamerican subsistence crops, but also specializing, in the Canyon area, in the cultivation of fruit trees in extensive irrigated orchards (cf. for example, Paso y Troncoso 1914). Agricultural activities were carried on in communally held lands cultivated by a free peasantry. Village cultivators were controlled by stewards who were responsible to local aristocratic magistrates, and minor rulers who themselves were dependent on the major rulers of the several Cuicatec city-states. These government officers were directly involved in the administration of agriculture.

Rulers did not control agriculture by monopolizing privately held land; the village communities were repositories of such rights. Rulers, however, were in charge of annual land redistribution among the peasantry. They directed corvee labor and controlled the processing of raw materials such as salt. The rulers supervised the redistribution of agricultural tribute for themselves and, in the period immediately prior to the Conquest, for the Aztec overlords who had conquered the region. They regulated the scheduling of religious festivities associated with the temples dedicated to water and fertility gods and goddesses. The rulers also controlled the sources of water for irrigation and the canals attached to individual towns and hamlets, and they exacted tribute (later taxation) from adjacent settlements which bought surplus water from their canal system. Such water tribute was collected on a seasonal basis. The Cuicatec ruling elite was not only a depository of the administration rights in land and water. They were also primarily responsible for directing defensive warfare against encroaching groups which in several cases attempted to obtain, by force, access to their water sources and improved irrigated lands (cf. example in Burgoa 1934). The ruling elite also relocated or founded communities to protect water sources. Moreover, in the sixteenth century, caciques, as their contemporary counterparts do today, manipulated water control as a source of power to obtain political following. For example, in a lengthy court case (1562), an illegitimate heir to the cacique post in the Cuicatec Estancia of Yepaltepec obtained public support of the Mazatec neighbors of an adjacent community by promising them that if appointed cacique, he would grant them Yepaltepec waters free. Up to the time, these villagers paid for the scarce water because they had lost a war several centuries before the Spanish Conquest over the rights to the water spring which fed their lands.

Caciques had a complex set of rights and duties vis-à-vis their subjects. There was a clear distinction between classes in terms of the division of labor and different usufruct rights over the means of production and surplus. Each class was also integrated into a social system which was clearly and markedly rank stratified, and which involved, among other power domains, differential control over water.

The maximum level of permanent political integration, however, did not occur at the level of the macro-ethnic group (defined by language), which formed only a loose polity integrated by symbiotic trade and intermarriage of ruling class members. We have discussed extensively in another work the possible causes of Cuicatec political fission above the city-state level ( E. Hunt 1972b). Each city-state, controlling its own independent water sources, was the maximum extension of the centralized political community. The separateness and isolation of scarce, small water sources gave the Cuicatec the characteristic Mesoamerican cantonalistic tendencies mentioned by Armillas (1949). But within each city-state, clear unification of authority existed, with marked differences in rights of water control and other political privileges between social classes. The functions of the elite were a monopoly, and acquisition of rights was primarily determined by ascription, by proof of aristocratic birth and rights of succession. Such rights were jealously guarded until the beginning of the seventeenth century, when the Cuicatec native elite gave way to pressure from the Spanish Colonists in the Canyon area.

Hence it is possible to argue that the native elite of the city-states controlled water because they had centralized political power in their hands, and at the same time that water control was one of the sources of their political power. However, the system of irrigation as it exists today in the Canyon and as it existed at the time of the Conquest has clearly discernible requirements related to the social need for technical and managerial skills and for unification of decision making to avoid conflict. Without effective control of the technological aspects, the irrigation network can physically collapse in a very short time. Without effective control of the sociological aspects, the irrigation system can fall into anomic internal conflict, which disrupts the normal flow of production and peaceful neighborly coexistence.

The Spanish Conquest produced just such a disruption of the normal Cuicatec political process and was accompanied by a steady lowering of production, reflected in the increased desperate petitions to reduce tribute on agricultural products. During the early period, many native irrigation works were abandoned, primarily perhaps because of the decimation of the population due to epidemics, but also because the social upheavals introduced by the Conquest disrupted the normal process of water control. Fights over water and irrigated lands erupted in the Canyon as well as the piedmont between villages sharing water sources. These were taken to Spanish courts in lengthy legal hassles. In the Canyon Spaniards soon found it profitable to displace the Indian peasantry and dedicate themselves to cultivating sugarcane and other irrigated products of interest to the Spanish entrepreneurs. Soon several major landed estates were formed (e.g., Tecomaxtlahuaca, Guendulain), some originally granted as encomiendas, some as private mills. These estates encroached on the best Indian lands and waters. The Spaniards built new irrigation networks to service the haciendas and mills. They granted rights on water to themselves at the expense of the Indian towns (this process is extremely well documented in the Ramo de Tierras of the AGN for the area north of the Cuicatec, in the lower Tehuacan valley).

Masonry and stone aqueducts soon dotted the Canyon countryside. During the seventeenth and eighteenth centuries these aqueducts were controlled either by the local Spanish Corregidores of the towns, or directly by the *hacendados* themselves, with support of the Spanish courts and colonial government. The small population size, a consequence of earlier disastrous epidemics, reduced land and water pressure in spite of extensive Spanish invasions. Drop in population density led to an "abundance of water" (Paso y Troncoso 1914, p. 187) for cultivation. This circumstance prevented major conflicts of access over water, judging by the absence of court cases.

The Colonial water works have been abandoned totally or are in a major state of disrepair, being either partially utilized in combination with open ditches, or left waterless as a monument to a bygone era when organized, major building activities of civil improvement were still possible.

In the nineteenth century, after independence, the colonial arrangements were hardly disrupted. Municipal officers controlled water for the whole Municipio of San Juan and throughout the district. But municipal authorities were often appointed and always were political dependents of the landed upper class. District chiefs (jefes politicos de distrito), the major local officials appointed by and directly responsible to state governors, were in San Juan always appointed from the families of major landowners. (Their grandchildren control San Juan today.) Some of the major haciendas in the Canyon were treated by the colonial and later independent government as separate political entities (municipalidades) and thus controlled their own water sources. Two small haciendas, however, were totally within the San Juan municipal boundaries, and could secure most of the water from the town's municipal system. One *hacendado* owned his own aqueduct within the municipio. There was thus a single official decision-making source for all of the users of any one municipal system. The *hacendado* class controlled district politics, including water politics, and used this control for their own benefit irrelevant of normative municipal regulations.

During the last period of the hacienda boom, in the late nineteenth and early twentieth centuries, population pressure on the Cañada led to expansion of hacienda control into the highland municipios, and to increasing unrest and conflict in the Canyon lands themselves. But the strong Porfirian government of the district maintained the elite control of water unchallenged. Changes in the social organization of irrigation occurred only after the upheaval of the Revolution. Bribing, uneven allocation of water, excessive water taxation of the poor, water theft, and other injustices in access to irrigation are in San Juan complaints which are no different today than they were in the late nineteenth century. However, the system functions fairly smoothly in spite of constant petty conflict, and there is no evidence to indicate that it may collapse. For half of the year (the rainy season), all fields receive sufficient water. In the dry season, the small cultivators either do not plant a second crop, plant a crop with low water requirements, or grow a stunted crop with insufficient water. They do not like it, but they cannot, at present, change the structure without another major revolution. The official monopolization of power over water by the states through the Water Commission, united with other sources of political strength on the part of the local ruling class, prevents both the reform of the irrigation system, and conflict running rampant. But even with the changes over a four hundred year period in the Cuicatec, water control has been, as far as our historical record shows, an important aspect of elite and local government prerogatives. The theoretical implications of this historical picture are discussed in the following sections.

## CONFLICT IN IRRIGATION

### Contemporary Internal Conflict in San Juan

One result of the complex picture presented above is that some conflict is generated by the contemporary contradiction between the normative expectations of the majority of water users and the actual distribution practices. The officially accepted egalitarian ideology with respect to access to irrigation water, which has been written into federal law, has been locally applied to the formation of bureaucratic mechanisms for water distribution. But the actual situation in San Juan reveals highly varied and competitive alternatives of control of land and water resources. The private control of land, water scarcity, the differential water demands of the various crops planted, the great differences in size and quality of fields, as well as the differences in wealth and power of the landowners involved, affect water allocations. The resulting uneven distribution of actual access to water, in contrast to the potential or ideal, results in constant petty minor conflict. This conflict in the past has occasionally developed into physical aggression and even homicide when the local government lost its grip on the reins of water control. Until about thirty years ago, the personnel in charge of water control were part of the municipal government staff. The water users had been, increasingly since the late nineteenth century, highly exasperated by the municipal handling of the town canal waters. The officers were arbitrary in their allocation of irrigation turns and the evaluation of water costs. They favored themselves and the town's largest and wealthiest landowners, demanding bribes and other unofficial costs. This control became more disruptive with the great increase in population size since 1900 and with the demands upon the water system for newly introduced, water-hungry cash crops such as rice.

Prior to the Mexican Revolution this situation has been accepted because the landed elite, in conjunction with its appointed local officers, had had the recurrent support of the state in their monopoly of both land and water. Then, however, because of favorable state attitudes, the town poor, who had had little voice over water control prior to the Revolution, found themselves supported by outside institutions which were hierarchically superior in official power to the local authorities.

During the immediate post-revolutionary period, conflict became rampant. The power of the local landed elite over water was openly challenged. *Usuarios* (water users) refused to pay water costs or participate in canal cleaning. Water stealing became a common practice and several violent fights and murders occurred in a period of a few years.

Once the traditional unified authority of the *hacendado* class and local rural elite was partially broken by land expropriation and other external events, conflict increased greatly. Many small cultivators who usually planted tomatoes in the grassland section discontinued the crop. Because tomatoes are a high water-demand crop, the uncertainties of obtaining water made their cultivation unprofitable. One informant said he completely stopped cultivating for several years, simply because there were too many dangers of becoming involved in fights over water or losing the whole crop because of water mismanagement.

To reestablish order, which was affecting small as well as large landowners to everyone's disadvantage, the community sought help by looking for external restraints in the shape of intervention by the federal government. A committee of water users petitioned the government to have a branch of the Federal Water Commission (*Junta de Aguas*) formed in town, and thus reduce the impact of increasing disorder. Many of the local elite (who controlled ditches for their private use) were opposed, but the majority won, and the Junta was created with the state's blessings. Its creation was an attempt to remove undue influences and resolve some of the past conflicts and alleged corruptions which interfered with an equal and just distribution of irrigation water. Decision-making power over water was therefore officially unified in a single agency, but also separated from other decision-making bodies which affected the whole community (especially the municipal government). In theory, control of water thus became unified, insulated from other political controls, and located above the municipal level. This agency was supposed to prevent interference of the locally influential power groups (the rural landed elite) in matters of water control, and to vest power in the small water users, who elected representatives from among themselves.

What in fact happened, however, is that the local elite now controls the Water Commission when necessary for their purposes, through their political clients who are the office-holders in that agency. This control is analogous to their control of the other formal government bodies in town. Many of the poor in San Juan now believe that the solution to the present problems of water allocation is to bring men in from the outside to manage the water system. Others believe that they should go back to the old system of municipal control, "without its corrupt elements." No one suggests, however, that the San Juaneros can do without some sort of unified control, nor can they suggest how to eliminate corruption or conflict except by increasing water supplies to meet current demands. One of our informants summarized their problem by stating flatly that "water makes people dishonest and peaceful men fight. If there is not enough, even the most honest man will be tempted to get an unfair share for himself and his friends. . .if he can find the way."[10]

## Inter-community Conflict

In a previous work we have extensively discussed the ecological, sociological, and political factors which hindered centralization of the Cuicatec above the level of the small city-states and their subject hamlets (E. Hunt 1972b). Many of these factors are not directly related to the economy of irrigated agriculture. However, the nature of the irrigation system may have been related, in part, to the schismogenetic tendencies of the communities of the district. First, the irrigation niches are small. Second, they served communities in which there was little if any colonial land pressure, and water was abundant given the size of the population of cultivators. Third, in most cases, a single city-state controlled a single major water source and its length. When the water source was formed by several minor affluent streams, a subject hamlet of the city-state was located in each headwater to protect it from enemy take-over or intrusion. If two subject hamlets shared a water source (e.g., San Lorenzo and San Francisco of Papalo), the ruler of the city-state administered the water and controlled the diversion canals and water allocation for both segments of the political community. Settlements of peoples who were not subject to the same city-state, located downstream (e.g., Quiotepec vs. Papalo), paid for the water in tribute. Legitimate rights to control of the length of a water course were reinforced by control of the spring from which a river emerged within the territory of the state, and by a history of military defeat of the downstream communities. Local, short-lived wars of defense of water sources which continued sporadically into the colonial period (under the name of *tumultos* or *montoneras*) reinforced traditional rights and reaffirmed the legitimate power of the city-state rulers to control their water sources. However, when a downstream community was superior in political might, population size, etc., to the upstream community which shared its waters, it appears that peaceful agreements between rulers were necessary about land and water. These agreements sometimes involved intermarriage of the ruler's successors.

The nature of the Cuicatec streams, however, makes the relation between upstream and downstream communities quite unusual. The Cuicatec streams are fed throughout their length by subterranean aquifers and cut the piedmont into deep, narrow canyons which traverse uninhabitable territory, making it impossible to use their waters for irrigation purposes in the middle section of their length. Thus, upstream communities cannot cut off access to water from downstream communities completely (although they have a marginal control over some water surplus). The communities of Cuicatlan and Papalo, which shared a stream until the present, exemplify this relationship. We do not know, however, how conflict over water during the dry season (if it existed) was handled between communities in the pre-Hispanic period.

We have argued that the evidence strongly suggests that soon after Conquest time there was no population pressure on the land, since chroniclers report that there was abundant water for all city-states reported. Thus, conflict did not arise in terms of water competition between Cuicatec centers, and pressures to centralize as a viable situation to avoid conflict were not present.

At present, several towns and municipalities engage every year in lengthy negotiations arbitrated by state officials over the quotas of water which they may share from common sources. The community of Quiotepec, which since remote times has shared a water source with Papalo, since these same times has been engaged in recurrent conflict with Papalo over water. In pre-Hispanic times they were controlled by different ethnic units (Mazatec vs. Cuicatec) and fought a war over the water rights. At the time of the Conquest, they were in constant open conflict about tributary rights which Papalo claimed over Quiotepec in exchange for water. This same conflict has been flaring up over and over again during the last 400 years, particularly in markedly dry years. The conflict does not reach armed attacks, however, because it is now handled through the state and federal courts, with regular state intervention and arbitration. When we were in the field, the male author of this paper was thoroughly welcomed and feasted by the villagers of Quiotepec, because they were under the erroneous impression that he was a secret agent of the government (an engineer) who was investigating Quiotepec's claims that they were being overcharged for water costs by Papalo, and forced to pay under the threat of having their available water reduced.

Arbitration of this type is common in Mexico. The community of Mixquic in the state of Mexico (a suburb of Mexico City which cultivates truck garden vegetables and flowers for the capital) has been involved in similar state-arbitrated water fights since colonial times.

Thus, at the village level, or within municipios, the local upper class united in their decision making with municipio-level government agencies which control water, unify decision making, and maintain peace between water users. Between villages and municipios, however, higher level, centralizing agencies deal with the allocation, arbitration, or introduction of reforms (both technical and distributional) in the irrigation arrangements.

Above San Juan, the district is administered by a bi-state watershed office of the Ministry of Hydraulic Resources. The administration is divided into several district watershed offices. Each is headed by a federal employee, usually an hydraulic engineer. Under him work several kinds of technicians: draftsmen, cartographers, architects specializing in designing waterworks, experts in irrigated agriculture, etc. For San Juan, the district covers two major rivers at the headwaters of the Aleman dam and Papaloapan River: the Grande and the Salado. Two states, Puebla and Oaxaca, are included in the district office. They provide services, advise towns and municipios on building new water works, facilitate government credit and supply free technical aid, aid in the reallocation of water schedules between conflicting towns, etc. This office cooperates with the water-control agencies of another ministry (Livestock and Agriculture) and with the district offices of the Administration of Communal Lands and Waters (Bienes Comunales). These agencies, operating above the municipal level, link the process of water control directly to the federal offices in the capital, and ultimately can be appealed only at the presidential level by petitioning the Secretariat of the Presidency. Final legitimization of any decision about irrigation water is therefore not decentralized. Neither communities nor municipios solve conflict by themselves in an arena without superior arbitration. All such local decisions, even if apparently made locally on the basis of consensus, are in fact subject to the final decision by higher level authorities in a pyramid of national political power. This pyramid, organized as a bureaucracy, has jurisdiction over water as well as over other national resources.

The historical evidence of the Cuicatec district firmly points to a single interpretation. When there is no pressure on the land or over water resources because of low population density, there is no water conflict either between communities or between neighbors in a community, nor is there pressure to unify the polity to insure peace. When water becomes scarce, because the number of cultivators and the land area under cultivation require more water than that which is available, conflict flares up. Water control under conditions of decentralization may be solved by internecine, short-lived warfare, or by political unification between the competing polities. In the sixteenth century in the Cuicatec, city-states within the language group solved minor conflict by intermarriage of the ruler's successors (see detailed account in E. Hunt, 1972b) and other amicable political agreements. Outside the language group, conflict was solved by either warfare or conquest and submission of downstream communities. But in the sixteenth century lack of land pressure optimized peaceful intervillage relations. Today, however, water control is regulated by the political machinery of the state, and under conditions of centralization, conflict is reduced to a minimum expression and officially sanctioned by state interference at the local levels in spite of great population pressure on land and water resources.

This situation, we believe, is common throughout Mexico, wherever conditions of water scarcity exist for canal irrigation, and where most irrigation occurs by utilizing small, permanent streams rather than major rivers which require extensive water works. In the areas where extensive water works have been built, both historically and in modern times, there is ample evidence that a clearly centralized political machine was in charge of the construction of the system, the allocation of labor resources, costs, and the decisions about allocation of resources.[11] It appears to be the case (as in El Shabana) that the physical size of linked or integrated water-control systems is related to the level of integration of the political community. The larger the polity, the more complex and extensively linked water works become possible. But these are not "necessary" when small water sources are used by a thinly settled population, and it is clear (from the Cuicatec case) that a limiting factor is the interaction between the potential of the environment, the technical resources of the society, and the population density.

## CONCLUSIONS

We wish to look at some of the results and consequences of our previous analysis in terms of the relationship between irrigation, power, conflict, and centralization. As soon as a physical irrigation system is sufficiently large, such as in Post Classic and contemporary San Juan, it seems to be inefficient. It is unlikely, and perhaps impossible, over the long run for a non-centralized system to work without great waste of energy in distracting conflict and anomic decision making. In San Juan, construction, but especially maintenance and policing, became concentrated in institutionalized political roles which include in function water control and differential access to power. In a system of this type, the maintenance of the major or core physical system is a labor-consuming undertaking. Normal repairs are a constant chore, and decisions have to be made every day about how much work is needed where. Then the work has to be supervised so that it is done properly and without damage. There are emergencies which require the ability to mobilize a substantial number of workers on short notice and to get them to do the work effectively.

In this arid land, with population pressure creating a water shortage, policing the system is a serious consideration, both in scheduling water allocations and protecting the receivers of the water; it is also necessary to detect theft of water and, most seriously, to resolve satisfactorily the conflict which arises over competition for a life-giving scarce resource. For a physical system such as San Juan, involving scores of domestic groups, routine administration even when thoroughly traditionalized and free of conflict as in Pul Eliya is complex and vital to the survival of the system. Bookkeeping for the system is of some importance also. The work has to be exchanged for money or other privileges, or the individual workers must be convinced or coerced into carrying out the work. The individual clients have to receive the amount of water which is regarded as their due and for which they may exchange either labor, tax, or tribute. Marked mismanagement of the administrative duties provokes open antagonisms. Lack of consensus has led to temporary anarchy, while recurrent injustices of distribution steadily build tension and conflict, to the point of social collapse. Coordination of decision making may function, however, to prevent such extreme developments.

If the system is large enough, it may encourage, if not require, that several functions be centralized by being allocated to specialized roles within the government apparatus: some of these roles require relatively scarce skills when compared with other roles in the same social system. Today even with the simple technological aspects of irrigation in San Juan, there is a differential distribution of knowledge, especially of engineering, organization and supervisory ability, and access to information. It is perhaps not logically necessary that all of these skills co-occur in the same role-syndromes, but all have to be available. It would appear that in San Juan skill at, and responsiblity for performing, the critical functions of the irrigation system (maintenance, allocation, conflict resolution) are not homogeneously distributed in the population. At the lowest skill level, all farm hands can perform the work of directing water into fields and cleaning the irrigation ditches. The more critical functions are those of deciding when to apply maintenance procedures, to effectively manage the maintenance crews, as well as allocation and the conflict it engenders in such a way that the system can continue to operate at some moderate level of productivity. These skills are far less widely manifested than the others, and tend to co-occur in the same role-syndromes monopolized by the upper classes. While these skills are by no means restricted to one or very few people in the social system (as witness the existence of four separate management and allocation agencies, two of which rotate persons through leadership roles for a small population), they are not widely distributed either.

Virtually all persons who occupy roles having to do with management of the irrigation system also occupy other political and non-political roles of control. That is, the roles which deal directly with irrigation are, for this system, thoroughly embedded in other political and economic networks. There are complex ways in which one activity system affects another. Skill in managing relations with the cash crop and money market are important in generating capital to manage and maintain the physical system. Maintaining a canal system is important for keeping satisfied political clients, and for keeping a profitable cash crop in production. All these interactions affect and are affected by skill in coping with outside bureaucracies, bribing, etc. Some of these interactions deal with water, but others do not, such as dealings with taxation officials.

The fact that the middle- and lower-class people could theoretically run the physical systems as well as the elite would seem to argue that the concentration of all functions in closed clusters of roles, and the linkage of these roles with other power roles in the social system, is not necessary from the point of view of agriculture. It is not so clear, however, that this degree of concentration of power is not necessary from the point of view of conflict resolution. The elite, who have a double function of control of the social aspect of the irrigation system and the local governmental apparatus, frequently involve themselves in internal matters of irrigation. It may well be, therefore, that the conflict resolution services of the elite are the most crucial to a situation of population pressure, which in part explains why the roles are so concentrated and monopolized in the very few persons at the top of the social structure.

While it is true that not all managers of water are elite, it is true that all politically powerful elite are or have been managers of irrigation. This fact suggests that irrigation is a vital part of the elite's power. It is not hard to see why it is so in San Juan. Irrigated agriculture yields cash crops, which generate cash. Cash is necessary for successful participation in the PRI, in private management of a canal system, and for the urban life style which the elite partake of. Irrigation also yields several opportunities for extensive patron-client ties. The more people who are dependent upon individual members of the elite's self-appointed management power, the more voters they have to support their faction in the local political contests. Also, their cash is crucial for the supply of credit to the local poor, a vital link in maintaining patron-client ties. In San Juan, therefore, water managers do not have to be elite, but elite have to be effective water-managers. The elite of San Juan need control of the irrigation system.

It is our hypothesis that the power elites are crucial for conflict resolution over water. The persons occupying water-control roles are intimately connected to the distribution of power in the society. (This connection is even true for the Sonjo, which is a simple, stateless society [Gray 1963].) It is certainly true for San Juan, in the sixteenth as well as the twentieth century, that water-control activities are intimately connected with power roles, which are built into the government apparatus in some way or other. These leadership roles are connected to land and to the higher levels of authority, have differential power with respect to water, and are stabilized in the local and national systems of stratification. It is clear, therefore, that power over water can be, and in this case is, linked with differentially distributed economic power, which in turn permits greater differential control over labor and capital for the construction and maintenance of the physical irrigation system. It is not theoretically imperative that such events occur, but our data suggest the hypothesis that this situation has a high adaptive potential.

We have, following Millon, advanced the thesis that a major issue in the consideration of irrigation systems is the degree of centralization of the conflict control system. In this instance, the question is how the San Juan social system copes with recurrent conflict between irrigators. At first glance the system seems to be centralized due to the existence of a single Water Commission. At second glance, it would appear to be locally decentralized, for there is not a single official unified decision-making body, but at least three such (the commissions, the municipal government, and the Ejido Commission), and in addition the mill and wealthier landowners who are also private canal owners exert pressures of their own. But if we look behind the scenes, control of conflict at least is hierarchically centralized. Conflict is kept under control (and this is the natives' belief as well as our observation) by the effective use of legitimate force employed by several local government agencies which are coordinated in their actions (e.g., the Water Commission and the local municipal police), and by the combined pressure of the cooperating wealthier landowners. Essentially, a few families which make up the San Juan upper class have unified ultimate decision making at the local level in their own hands by effectively controlling all dimensions of the body politic. This situation has existed since the sixteenth century, in spite of considerable radical changes in population size, crops planted, governmental organization at the local level and the larger political system, and periods of great political change and reform at higher levels of the state structure.

The major question to which we addressed ourselves in this paper is the relationship between generalized political centralization and levels of conflict in a society based on canal irrigation. In more general terms, this question focuses on the problem of the evolutionary adaptive value of centralization in societies with irrigated agriculture. The hypothesis which emerges from our case study is that one condition under which centralization of authority is adaptive in reducing conflict is under conditions of water scarcity (i.e., when there is population pressure on land and water resources) and that this adaptive response might be particularly effective in a system of food production which is totally dependent upon canal irrigation agriculture.

We suggest in this paper that without unified decision making, high levels of social disruption and conflict will tend to paralyze the socio-economic system. Over long periods of time, the threat of anarchic conflict will generate further pressure to centralize the system, if it is to survive. However, stating that a variable is logically or functionally adaptive, or it has evolutionary potential, does not necessarily imply that it will always be adopted. Secondly, centralization of decision making over water is probably not adaptive under all conditions, and it is most likely adaptive only up to the maximum territorial extension of the particular irrigation system. We suggest, hence, that when water becomes a socially scarce resource, control of canal irrigation, e.g., the control of springs, water allocation, etc., can be manipulated to diminish or increase conflict for political purposes within the political system in which the irrigation system is imbedded.

Centralization of the political system facilitates higher levels of productivity by reducing social conflict at the local levels. It also may allow the formation of an institutional frame and specialized roles which could be potential springboards to other realms of local power or control, if the social structure allows for a stratified control of resources. However, in all probability, there is a feedback loop involved here because sustained power in other realms of the society will probably increase the likelihood of particular persons either assuming themselves or being regularly recruited to fill the decision making roles connected with irrigation. At this time, we see nothing in the data which permits us to legitimately decide to make irrigation either the independent or dependent variable, when considered with other kinds of politically controlled systems within the same society. But neither can we suggest from the empirical evidence that the irrigation system is epiphenomenal to the political process or to the degrees of centralization of authority which may be most adaptive to maintain social conflict at low levels. On the contrary, irrigation appears to be, for the system considered here, one of the major foci of the political field.

## NOTES

[1] The field work and archival research on which this paper is based was generously supported by NSF Grant GS87; the major writing was produced when the senior author was freed from teaching duties by NSF Grant GS3000. We wish to thank Professors Gibson and Downing for first suggesting the participa-

tion of E. Hunt in the Long Beach symposium. The senior author also wishes to thank the faculty and student body of the Department of Anthropology of the University of Arizona, who gave her the opportunity to present a preliminary version of this paper, and who generously offered useful comments. Robert McC. Adams, Robert Betteral, George Cowgill, Jeremy Sobloff, and Alex Weingrod read a draft and offered suggestions. We alone of course are responsible for its final form. The ethnographic present refers to the years 1963 to 1966. The bulk of the data was gathered in 1964.

[2]A partial exception is the trend among archaeologists to probe into the evolutionary potential of irrigation vis-à-vis the state, which was not a part of Wittfogel's original conception.

[3]This method will lead, at times, to what might appear to be a surfeit of ethnographic minutiae, but we prefer to differ from past accounts by swinging to the side of over-documentation.

[4]The nature of the latent existence of national, centralized power and its infrequent manifestation has important implications for field research strategies. It is a difficult topic to gather data on, precisely because the events are infrequent and apparently confused with other kinds of institutions such as stores, land transfer, and political parties. Several years may go by without a case becoming public. Moreover, it is a subject on which informants are reluctant to talk at all. At one point during field work, we were prompted to conclude that the elite of San Juan showed as much willingness to talk about water as large American businesses show about discussing price fixing and monopolies. The first response of most informants in San Juan, even those who had little if any personal concern with water, was "that never happens here," "we never fight about water," "here everything is pacific." It takes further probing to uncover the fear of retaliation by the local elite (if they are caught speaking) or of "getting into trouble with the authorities." This phenomenon is not common only to San Juan. Lees (1970), who was able to successfully administer a formal survey schedule on irrigation in Oaxaca, found great difficulty in gaining access to information which dealt with actual practices, such as lists of irrigators, and watering schedules.

We suggest that archives, especially the court archives, are excellent sources of particular cases, and should be looked into. Diligent inspection of a few cases should be enough to establish promising hypotheses about the patterns for resolving such disputes, which can then be investigated with informants.

[5]Figures for volume of water for the minor streams is lacking. Only the major rivers (such as the Grande) have been investigated by Mexican hydrologists and geographers. The text above is based on strong impressions, sporadic knowledge, and educated guesses of the local head of the district watershed, who is a hydraulic engineer specializing in the construction of diversion dams. We are grateful for his patient assistance in answering our naive question on the topic of water.

[6]The total land under irrigation controlled by San Juan is a little more than one thousand hectares. Our figures are incomplete. We have both official and unofficial figures for Spanishland, the privately controlled canals, and the ejido. These come from the Ministry of Hacienda, canal owners, mill administration, and other informants. Figures for the grass land were obtained in maquilas (a measure of seed used in planting) and transformed into hectares. However, different fields require different amounts of seed. The average is between 1.5 and 2 maquilas per hectare. Unfortunately, we were not able to obtain official figures for the grass land because at the time of field work the office of the tax inspector dealing with registered lands was involved in a complicated tax dispute with the town's elite (particularly the faction in power at the time), and we found that our attempts to obtain information were in danger of being misinterpreted as interference with the conflict.

[7]Prices are presented in the ethnographic present of 1964.

[8]Other factors are the disruption of sleep and work schedules, and the supernatural dangers associated with the night.

[9]An extensive treatment of these materials can be found in E. Hunt 1972b. All relevant published and unpublished evidence is presented there, accompanied by an annotated bibliography. Reasons of space preclude repetition here.

[10]"El agua hace a la gente deshonesta y un peleador al hombre tranquilo. Si no hay bastante, hasta el mas honesto trata de agarrar más de lo que le toca para él y sus amigos. . .si encuentra el modo."

[11]Examples are the Netzahualcoyotl dike in the Tenochtitlan Lake and such contemporary works as the Aleman dam in Vera Cruz.

## REFERENCES

ADAMS, Richard N.
1971 *Crucifixion by Power.* New York: Random House.

ADAMS, Robert McC.
1966 *The Evolution of Urban Society. Early Mesopotamia and Prehispanic Mexico.* Chicago: Aldine Press.

ARMILLAS, Pedro
1949 *Notas sobre sistemas de cultivo en Mesoamerica: cultivos de riego y humedad en la Cuenca del Rio de las Balsas.* INAH, Anales, 3:83-113.

ATTOLINI, José
1949 *Economia de la cuenca del Papaloapan: Agricultura.* Mexico: Instituto de Investigaciones Economicas.
1950 *Economia de la cuenca del Papaloapan: Bosques, Fauna, Pesca, Ganaderia e Industria.* Mexico: Instituto de Investigaciones Economicas.

BURGOA, Fra. Francisco de
1934 *Geografica Descripcion Publicaciones del Archivo General de la Nacion,* XXV. Mexico.

CASO, Alfonso
1953 *El Pueblo del Sol.* Mexico: Fondo de Cultura Economica.

COMISION DEL PAPALOAPAN
1956 *Atlas Climatologico e Hidrologico de la cuenca del Papaloapan.* Estudios y Proyectos, A.C. Mexico.

FERNEA, Robert
1970 *Shaykh and Effendi: Changing Patterns of Authority Among the El Shakana of Southern Iraq.* Cambridge: Harvard Univ. Press.

FLORENTINE CODEX
1950 Bernadino de Sahagun. *General History of the Things of New Spain.* Anderson and Dibble, (trans. and eds.) Santa Fe: Monographs of the School of American Research, #14.

GEERTZ, Clifford
1963 *Princes and Peddlers.* Chicago: University of Chicago Press.

GLICK, Thomas
1970 *Irrigation and Society in Medieval Valencia.* Cambridge: Harvard Univ. Press.

GONZALEZ CASANOVA, Pablo
1965 *La Democracia en Mexico.* Mexico: Ediciones Era.

GRAY, Robert
1963 *The Sonjo of Tanganyika: An Anthropological Study of an Irrigation-based Society.* International African Institute. London: Oxford Univ. Press.

HOPKINS, Joseph
1970 *Report of an Archaeological Survey to the Instituto de Antropologia.* Mexico.

HUNT, Eva
1969 *The Meaning of Kinship in San Juan: Genealogical and Social Models.* Ethnology, 8:37-53.
1972a *Irrigation and Centralization: A Critique of Millon's Argument.* Paper read at the Annual Meeting of the SWAA.
1972b *Irrigation and the Socio-Political Organization of the Cuicatec Cacicazgos.* F. Johnson and R. MacNeish (eds.) The Prehistory of the Tehuacan Valley, Vol. IV. Austin: Univ. of Texas Press. (Monograph)

HUNT, Eva and Hunt, Robert
1969 *The Role of Courts in Central Mexico.* In Bock (ed.), *Peasants in the Modern World.* pp. 109-139. Albuquerque: Univ. of New Mexico Press.

HUNT, Robert

1965 *The Developmental Cycle of the Family Business in Rural Mexico.* In J. Helm (ed.), *Essays in Economic Anthropology,* Proc. 1965 Ann. Spr. Meeting, AES, pp. 54-79. Seattle: Univ. of Washington Press.

1971 *Components of Relationships in the Family: A Mexican Village.* In F. L. Hsu (ed.), *Kinship and Culture.* Chicago: Aldine Press.

LEACH, Edmund

1961 *Pul Eliya.* Cambridge: Cambridge University Press.

LEES, Susan

1970 *Socio-Political Aspects of Canal Irrigation in the Valley of Oaxaca, Mexico.* Unpublished Ph.D. Thesis, Univ. of Michigan.

MARTINEZ GRACIDA, Manuel

1883 *Cuadros Sinopticos. . . de Oaxaca.* Anexo 50 a la Memoria Administrativa Presentada al Congreso del Mismo el 17 de Diciembre de 1883. Oaxaca: Imprenta del Estado.

MILLON, René

1955 Trade, Tree Cultivation and the Development of Private Property in Land. *American Anthropologist* 57(4):698-712.

1962 Variations in Social Responses to the Practice of Irrigation Agriculture, pp. 56-88 in R. Woodbury (ed.), *Civilizations in Desert Lands.* Anthro. Papers #62. Dept. Anthropology, Univ. of Utah.

PADGETT, L. Vincent

1966 *The Mexican Political System.* Boston: Houghton Mifflin.

PASO Y TRONCOSO, F.

1914 *Papeles de la Nueva España: Relaciones Geograficas del Siglo Dieciseis.* Segunda Serie, Vol. IV.

PAZ, Octavio

1972 *Three Faces of Mexico.* Paper read at a Meeting of the Brandeis University Latin American Committee.

PIÑA CHAN, Roman

1960 *Mesoamerica.* INAH, Memorias No. VI. Mexico.

POLEMAN, T. T.

1964 *The Papaloapan Project: Agricultural Development in the Mexican Tropics.* Stanford: Stanford Univ. Press.

PRICE, Barbara

1971 Prehispanic Irrigation Agriculture in Nuclear America. *Latin American Research Review,* 6:3:3-60.

RELACION DE CUICATLAN

1914 Escrita por Juan Gallego. In Paso y Troncoso, Francisco (ed.). *Papeles de la Nueva España,* Segunda Serie, Vol IV: 183-89.

SAHLINS, Marshall

1962 *Moala.* Ann Arbor: Univ. of Michigan Press.

SIMPSON, Leslie Byrd

1937 *The Ejido: Mexico's Way Out.* Chapel Hill: Univ. of North Carolina Press.

STEWARD, Julian, ed.

1955 *Irrigation Civilizations: A Comparative Study.* Washington, D.C.: Pan American Union Social Science Monograph, No. 1.

WITTFOGEL, Karl

1957 *Oriental Despotism: A Comparative Study of Total Power.* New Haven: Yale Univ. Press.

1972 The Hydraulic Approach to Pre-Hispanic Mesoamerica in Richard S. McNeish, et. al. *The Prehistory of the Tehuacan Valley,* Vol. IV:59-80. Austin: Univ. of Texas Press.

CHAPTER 13

# IRRIGATION DEVELOPMENT AND POPULATION PRESSURE

Wayne Kappel

*Department of Anthropology, Temple University*

In this paper[1] attention is focused on a few of the pressures and accommodations involving irrigation which may contribute to a succession of organizational types.[2] The aim is to demonstrate mutual relationships between the growth of population, the growth of irrigation systems, and the nature of decision-making groups which manage irrigation. A review of the relevant literature is presented, followed by criteria for measuring irrigation size, population size, and centralization of decision making. Data from seventeen societies, arranged on an ordinal scale, are used to imply relationships between the variables. Finally, some speculations are offered about the sequential order of the relationships.

It is assumed that any persistent human group attempts to maintain its integrity "through the operation of adjustive mechanisms. . .to environmental pressures" (Alland 1967:120). Any significant interference with the environment will require adjustments throughout the system (Sahlins 1964:132, 137). The adjustments which are precipitated and maintained by human decisions via positive feedback from the interference (Flannery 1968:68) may contribute to the short-run survival of a society. From the point of view of long-run continuity, the decisions are sometimes successful, sometimes not (cf. Bennett 1966:33).

The hypotheses developed here rest upon assumptions derived from Boserup (1965:11) and Harner (1970:68-71): Population growth increases resource scarcity,[3] and increased scarcity brings about consequent socioculture structure and change.

## HISTORY OF THE PROBLEM

Of the chosen variables, irrigation has been most closely associated with decision making in the form of political authority type. The relationship was outlined by Wittfogel in 1938 and elaborated in *Oriental Despotism* (1967). Wittfogel (1957:344) has summarized the main points of his position in the following way:

> Hydraulic society is a special type of agrarian society. Its peculiarities rest on five major conditions:
>
> (1) Cultural: the knowledge of agriculture.
> (2) Environmental: aridity or semi-aridity and accessible sources of sources of water supply, primarily rivers, which may be utilized to grow rewarding crops, especially cereals, in a water deficient landscape. A humid area in which edible aquatic plants, especially rice, can be grown is a variant of this environmental pattern.
> (3) Organizational: large-scale co-operation.
> (4) Political: the organizational apparatus of the hydraulic order is either initiated or quickly taken over, by the leaders of the commonwealth who direct its vital external and internal activities – military defense and maintenance of peace and order.
> (5) Social: stratification separating the men of the hydraulic government from the mass of the 'people.' The rise of a professional, full-time bureaucracy distinguishes primitive hydraulic society. . .from the state-centered forms of hydraulic societies. . . .
>
> The hydraulic type of agrarian society was not confined to China. Historical evidence indicates that agrarian civilizations with government-directed water control originated several thousand years before the Christian era in the in the Near East, in Egypt, and in Mesopotamia. Similarly structured societies emerged early in India, Persia, Central Asia (Turkestan), many parts of Southeast Asia, and in Java, Bali, and ancient Hawaii.
>
> In the Western hemisphere such societies flourished prior to the Spanish conquest in the Andean zone (culminating there in the great irrigation empire of the Incas), in Mesoamerica (particularly in the region of the Lake of Mexico), and in the southwestern United States in Arizona (Hohokam) and, on a tribal scale, in New Mexico among the Pueblo Indians.

Essentially Wittfogel asserts that in order for agrarian society to exploit a dry environment, the necessity for larger-scale hydraulic works develops.

The organization of the hydraulic works influences the structure of the government and other institutions, causing the socio-political type called "Oriental Despotism" to emerge. That is to say, "the necessity of hydraulic works gave rise to managerial administrative control, on the basis of which despotic institutions and control developed" (Eisenstadt 1958:436). The political and social processes in despotic societies embody an absolute ruler "who rules according to his own wishes and interests. . .[and whose] rule is not checked by 'laws of nature or cultural traditions' or by the strength of any group in the society". The underlying thesis is that "the state is stronger than the society" (Wittfogel 1967:50). Wittfogel's proposition has been accepted by some anthropologists and historians as an explanation for the origin and maintenance of civilizations. Others have criticized his use of a linear-causality model to connect irrigation and the nature of political structure and process. Furthermore, the inapplicability of Wittfogel's interpretation to a number of historic cases has been frequently emphasized.[4]

Steward has proposed ". . .a developmental typology not only for the irrigation areas of the Central Andes but also for Mesoamerica and similar Old World areas" (1955: 18, 94, 194, 203-5). He is generally in agreement with Wittfogel's attempt to "use historical data to make functional-historical analyses of the socioeconomic structure of early civilizations," but Steward reserves judgement about the exact mechanism by which irrigation may be related to political authority in all cases. He recognizes that other variables – population density, amount of surplus, and extent of trade – affect the degree of absolute power concentrated in the political or religious authority. Thus, he reserves his final interpretation of Mesoamerican civilization until more data are available.

Robert McC. Adams (1968:370-71) distinguishes between large- and small-scale irrigation systems: small-scale irrigation "does not represent a heavy investment in labor. . .[and its] construction and maintenance. . .requires no elaborate social organization." Although noting that large-scale irrigation and bureaucratic authority are found to coincide in four historic civilizations, he argues that ". . .the introduction of great irrigation networks was more a 'consequence' than a 'cause' of the appearance of dynastic state organization" (cf. Carneiro 1970:738 quoting Gernet). The similarity in these four civilizations is seen as grounds for the creation of a "type" of society which had arisen in a variety of habitational zones. Surpluses made available through sedentary, diversified, intensive agriculture provide other typological criteria. Like V. Gordon Childe (1962), Adams (1965:40-48) sees surplus providing leisure time through which increased social stratification may come about. Harris (1959:195) holds a similar position, but adds that a significant distinction must be made between individual surpluses and social surpluses if we are to interpret their relationship to social factors. In the same vein, Jacobs and Stern (1947:125-26) contrast the functional difference between small and large surpluses, stating that "the amount of surplus produced is the crucial determinant of other aspects of the economy and social system of the people. It plays a significant part in deciding the density of the population and the complexity of social organization." To surplus, Boulding (1962:162) adds "politics, that is, exploitation" as an ingredient for creating civilization.

Similarly, Hole's (1966:609-10) model presents an interplay of economic surplus and emergent centralized political authority in Mesopotamia. Drawing upon Sahlins' work in Polynesia, he states that "the greater the surplus is, the greater is the degree of stratification." He mentions the "contributory" factors of urban centers and irrigation for political centralization, and adds that "an irrigation system also implies that some men may have more direct control over the supply of water than others." However, Netting, Spooner and Downing (in this volume) indicate that there is no necessary causal relation between irrigation and increasing power of *individuals.*

As far as surplus is concerned, Orans (1966:30) has questioned whether it is really the primary variable which leads to the elaboration of social stratification. He suggests, rather, that ". . .an increase in population might contribute to increased stratification." Harner (1970:84-85), following a similar line of reasoning, is convinced that a scarcity model is more useful than surplus or energy growth concepts for predicting social evolution.

Questioning not just elements of the Wittfogel hypothesis, but the entire formulation, Millon (1962:80, 86) concludes that no direct relationship exists between population size, centralization, and irrigation. His findings have been questioned by Sanders and Price (1968:81, 181-83, 198-99, cf. 176) on the basis of methodology.[5] Although agreeing in part with Millon that "no absolute relationship exists between the size of a small irrigation system and the degree of centralization of authority in its utilization," Sanders and Price construct a scheme for large-scale works that is similar to Wittfogel's. Differences in particular regional arrangements ". . .do not negate the principle that a functional relationship exists between centralized power and relatively large irrigation systems, or that the former is an effective solution to the problem of the operation of the latter." "Conflict [over the allocation of increasingly scarce water] stimulates the selective process in favor of centralization of authority – the more severe the conflict, the greater the need for and probable evolution of centralized control." Increased sedentariness and diversity of intensive agricultural exploitation yield a surplus. A surplus brings about the beginning of specialization and allows for a crucial population increase. "If population growth is one of the major achievements of the evolution of ecological systems, then the development of centralized control of irrigation systems can be understood."

More than a decade ago, Eric Wolf (1959:58-61, 81) attributed the beginnings of Mesoamerican civilization to an essentially environmental variable: the availability of relatively fertile and flat lands in highland Mexico. Agriculture prompted the development of irrigation systems in these areas. A theocracy was necessary to administer the irrigation technology and to distribute agricultural surplus. In an earlier study Wolf and Palerm (1955:280) wrote that "once established, of course, irrigation probably operated in turn to centralize and intensify political controls," but was not a necessary cause. On the contrary, they held, as does Adams, that the centralized authority of the state brought about large-scale irrigation as an accommodation to increased population. Further, Palerm (1955:29-30) suggested "the direct relation between population density and the agricultural system" was important in distinguishing small-scale irrigation from larger works.

Although "social scientists have long been aware that a relationship exists between the size of a society in terms of population and its degree of complexity. . .no cultural anthropologist has attempted to determine this relationship precisely" (Carneiro 1967:234, 239). Using a Guttman scale, Carneiro orders 354 traits including population size. He concludes that "a more complex social structure [including increased centralization of political authority] does not arise *solely* as a response to increased population"; the demands of subsistence or defense may also bring about increased complexity. An alternate response to increased population or economic and military demands is "simply to split" the population into more scattered settlements. However, ". . . if a society does increase in size and if it remains unified and integrated it must elaborate its organization." The mechanism by which bounded agricultural areas bring about this elaboration has been labeled "circumscription" by Carneiro (1970).

Implicit in Carneiro's theory is population increase, either through natural growth or immigration. In several of the formulations outlined above, population pressure is included as a contributing factor, but is seen as dependent upon surplus. Ester Boserup, in a revolutionary book (1965), shows that taking population increase as the independent variable leads to more productive research. Simply stated, her thesis is that intensification of agricultural techniques, including irrigation, is a direct result of population pressure. A population-growth model requires time-depth and would seem eminently applicable to the development of ancient civilizations. However, there is a problem in deriving figures for population size in antiquity. Archeologists have attempted estimates of population by inference from the size of ancient sites, monuments, and number of burials. Kaplan (1963:390-99, 401) has reviewed with skepticism the population figures and the allusions to a centralized, elite, "state-like" polity proposed by Sanders, Mayer-Oakes, Palerm, and others. He doubts the "relationship between the construction and maintenance of large-scale public works and the emergence of centralized bureaucratic political systems." In Mesoamerica, "the characteristic mode of constructing large-scale public works was in discontinuous stages. . .over long periods of time." In response to Millon's (1960b:379) statement that "the Pyramid of the Sun in its present form was constructed in a single building operation," Kaplan (1963:401) asks, "What does 'built in one operation' mean in terms of time – 25 years, 50 years, 100 years, 200 years?"[6] Kaplan insists we must know precisely the number of man-hours over time that were expended in building large-scale public works.[7] He (1963:404-5) cites Needham regarding the similarity in scale between the water works of ancient Ceylon and China and remarks that the history of Ceylon ". . .reveals little trace of a complex bureaucratic organization." Leach (1959:7-14) supplies affirmative data for Kaplan's statement. The monumental structures of ancient Ceylon were built discontinuously over long periods (hundreds and even over a thousand years) without massive labor forces and in the absence of a centralized bureaucratic state.

Eisenstadt (1958:440, cf. Eberhard 1958:447), in a review of Wittfogel, writes: "In many 'hydraulic' societies the management of hydraulic works is not necessarily always in the hands of the central bureaucracy. Eberhard has shown that in China many of those works were in the hands of the local gentry."

The application of Wittfogel's thesis to the southwestern United States has met with similar objections (Woodbury 1961:550-53): "There is little Hohokam material to suggest [a centralized authority] although a few imported luxury items from Mexico, a few distinctive burials and a few architectural details. . ." are found. Woodbury (1960:260-70) shows that the Hohokam irrigation canals in use at any point in time were small, and later concludes that "it is more probable that the canals were built with neither a central authority. . .nor a large labor force. They were probably built gradually, and a small ditch of two or three miles was repeatedly enlarged through several generations until final size was reached. . . . It was far removed from even an incipient 'Oriental Despotism' " (1961:556-57).

## THE PROBLEM

The relationship between large-scale irrigation works and the development of centralized political authority still remains unclear. The most frequently mentioned variables suggested to influence this association are: (1) the size and the density of the population, (2) the size of the irrigation facilities, and (3) the degree of centralization of the decision-making political authority. There are six possible ways to order these variables in linear, implicational statements:

a. population → irrigation → political authority
b. population → political authority → irrigation
c. political authority → population → irrigation
d. political authority → irrigation → population
e. irrigation → political authority → population
f. irrigation → population → political authority

Almost all of the six orderings have been suggested at one time or another. The remainder of this paper will attempt to determine which of these sequences occur by examining information drawn from seventeen ethnographic cases.

### Definition of Variables

*Extent of Irrigation.* The size of irrigation systems varies greatly and is difficult to appraise. The scale of irrigation system uses as criteria: (1) the miles of canals in use at one time, (2) the number of acres served at one time and (3) statements by the ethnographers as to the relative size of the system in that order.

*Effective Population Density.* Population may be measured absolutely or in terms of density. For our purposes the latter is preferable because at any given level of demand on resources, an increase in density (not necessarily in absolute number) exerts exaggerated pressure on limited resources (Harner 1970:68; Zubrow 1970:2-4). Beardsley *et al* used the "community" as their unit of study, which is not always the economically functional unit in terms of resource exploitation; therefore, for this paper it is an inappropriate unit. Population densities were determined by the following criteria in order of availability: (1) reported densities, (2) mean densities computed from population and geographic area figures, (3) ethnographer's judgement (approximate figures to support adjectives such as "dense," "sparse," etc.) and (4) types of settlement reported ("dispersed," "small village") in conjunction with the number of people reported in such settlements. The computed means in all cases are population in relation to total land, not just cultivated land.

*Centralization of Political Decision Making.* This variable was the most difficult to quantify. The final criteria come from Weber's definition of Traditional Political Authority Type (Gerth and Mills 1958). Political authority is considered more centralized if: (1) The number of individuals who contribute to making the decision decreases, (2) the succession to authority positions becomes more restrictive, (3) the legitimization of authority and extensive coercion become grounded in unchallengeable tradition and (4) the autonomy of initiating decisions and resorting to their enforcement is carried out.

### Relations of Variables

Originally, I intended to draw from Murdock's *Outline of World Cultures* (1957), a stratified random sample which would be minimally affected by the possibility of historical interconnection and its biasing effect on a functional association (or Galton's problem). A large enough sample would be amenable to statistical treatment, and probability statements could then be made. However, the corresponding data in the HRAF for the randomly selected cases were not detailed enough to allow for quantification of the variables according to the established criteria above. Therefore a judgement sample of seventeen cases was selected. Each case has irrigation and included sufficient detailed information about the other variables. Additional cases have been called to my attention and are currently being added to create a sample large enough to treat statistically. The cultures used in this paper are:

1. Sabi Valley, Pakistan
2. Apa Tani, Burma
3. Swazi, South Africa
4. Colorado High Plain, USA

5. Chake, Maracaibo
6. Kurds (Zagros Mts.), Northern Turkey
7. Kachin, Burma
8. Aracay, Maracaibo
9. Magar, Nepal
10. Kond, Iraq
11. Sivah, Libya
12. Arawak, Northern Venezuela
13. Kharia, India
14. Caqueto and Jirijara, Maracaibo
15. Khosi, Assam, India
16. Pathan, Pakistan
17. Araucanians, Chile

The variables for each society (consisting of the series of criteria given above) were plotted. In making the graphs, the problem of obtaining comparable data recurred. This problem is especially evident for the population data; the use of both size and density were required, e.g. five persons per acre, but also 5,000 persons total. The descriptions found for the irrigation systems and the decision-making systems were somewhat more conducive to classification. The various descriptions were reduced to four categories for these two variables.

These findings do not permit speculation about the origins of irrigation or population increase. However, something can be said about how the variables are related. As population increases to about five persons per acre, both irrigation and centralization increase in scale. At five persons per acre, irrigation systems are no longer independent family enterprises operated by the household, but village projects administered by a village council. Although the village irrigation system remains the autonomous functional unit, when population increases from 5 persons per acre to 5000 persons in the total unit, decisions about these systems are no longer made by a council but by one official. Irrigation systems of district size are characteristic of populations of more than 5000. When populations exceed 5000, decision making is held by the office of a non-indigenous (government) functionary instead of one local official.[8] The "lag" between scalar increase in centralization of authority and increase in irrigation system size may be interpreted to mean that under conditions of population growth, investments in centralization of management are made before investments in the elaboration of irrigation. These interpretations would

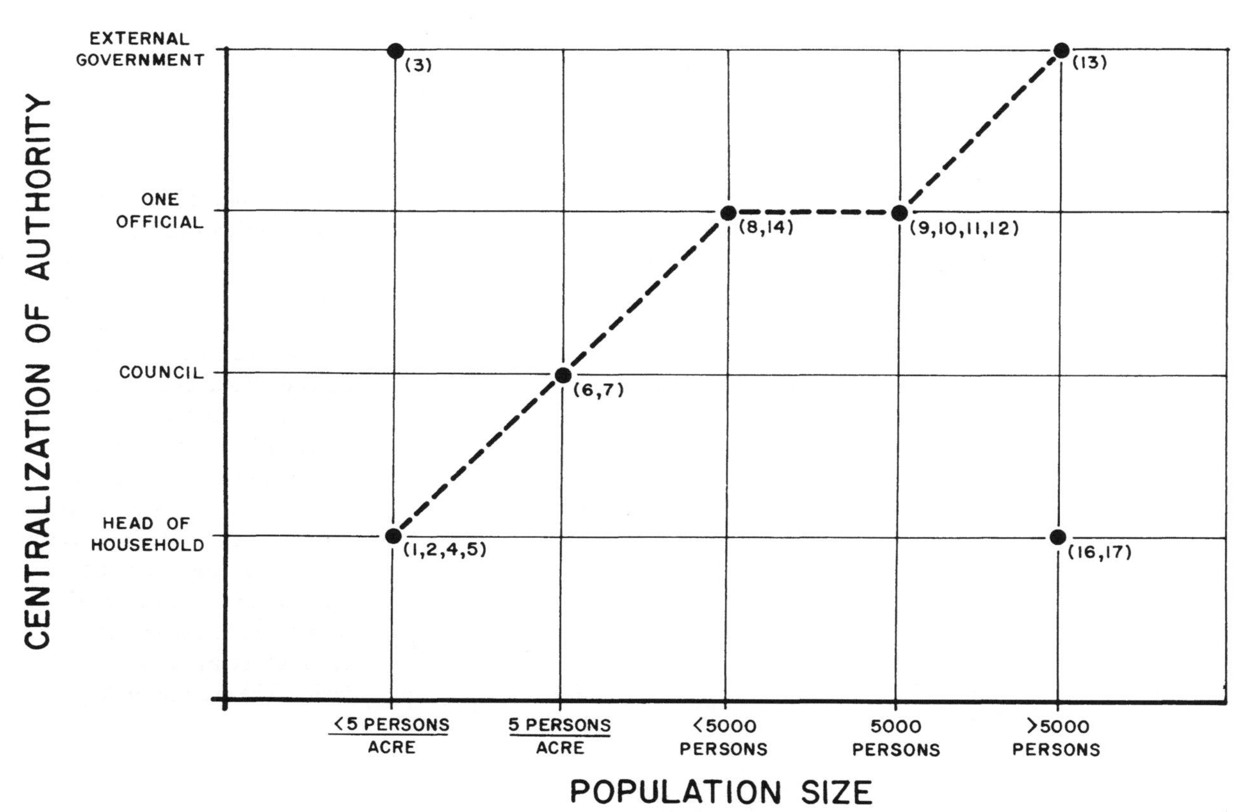


Fig. 13.1 Relationship of centralization of authority and population size for seventeen societies. Source: opportunistic sample of 17 cases from HRAF.

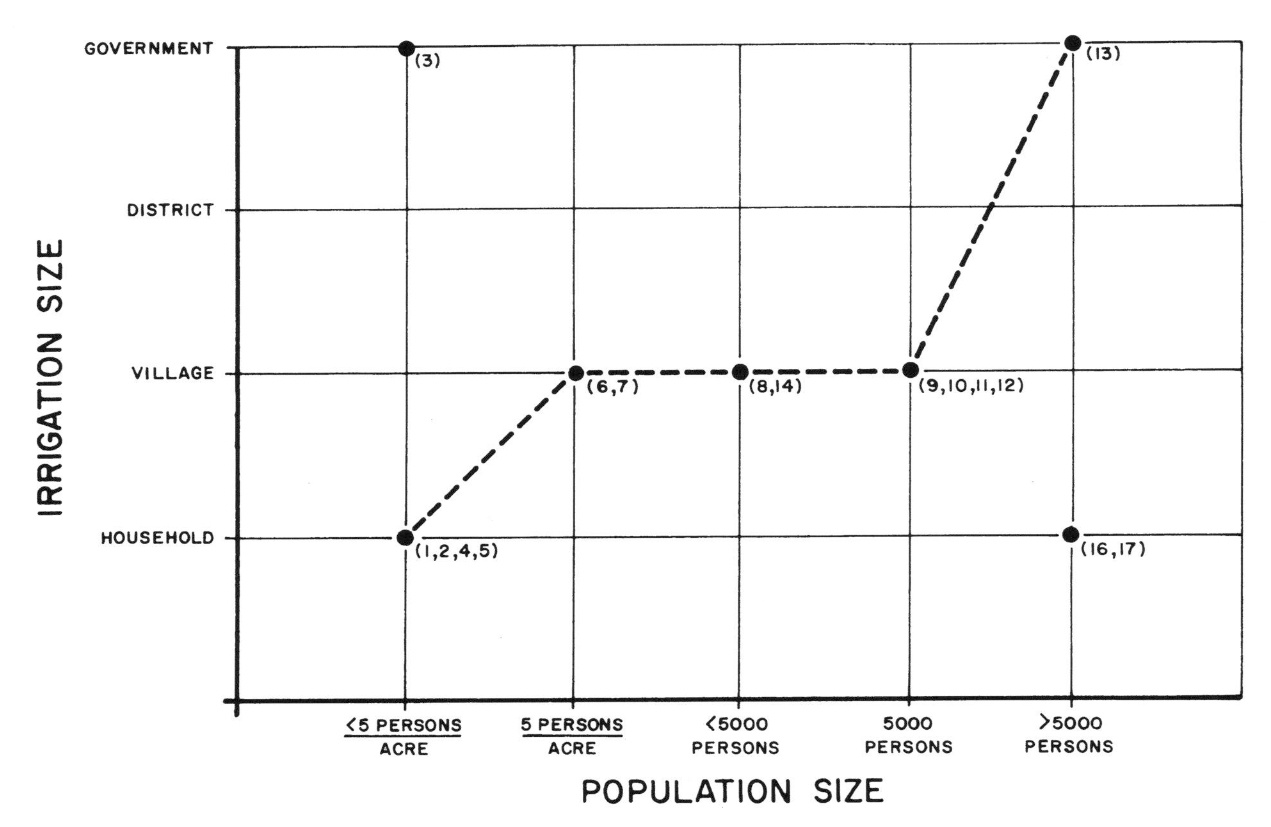


Fig. 13.2 Relationship of size of irrigation and population size for seventeen societies. Source: opportunistic sample of 17 cases from HRAF.

seem to support those who have argued for viewing population pressure as an independent variable in social change. Additionally they indicate that either implicational statement a or b above is plausible for populations under 5 persons per acre, but that statement b accounts for ordering the variables for populations increasing beyond five persons per acre. Thus, the graphs illustrate the contention of Adams (1965), Carneiro (1970), and others that major changes in irrigation system size occur only after changes toward centralization have already been made.

## NOTES

[1] My thanks are extended to several people for contributing their time and skills to this paper. I profited from discussions with John W. Bennett about an early and somewhat different version of this paper. T. Patrick Culbert, William A. Longacre, William L. Rathje, George Esber and Stephen A. Kowalewski gave helpful criticisms to subsequent drafts. Theodore E. Downing supplied valuable editorial and organizational suggestions which I have followed. I am especially indebted to McGuire Gibson for many hours of discussion and help in writing final draft revisions.

[2] I am reluctant to use the term "evolution" to describe the kind of changes that this paper suggests. For that reason I have borrowed the term "succession" from field biology. Ecological succession refers to the "... orderly and progressive replacement of one community by another until a relatively stable community occupies the area. Each seral stage is a community, although temporary, with its own characteristics, and it may remain for a very short time or for many years. Some stages may be completely missed or by-passed" (Smith 1966:127-28). Abstracting some common trends of succession, Margalef (1968:30-31) says

> biomass increases during succession as, almost always, does primary production; however, the ratio of primary production to total biomass drops. Diversity

very often increases. Sometimes diversity increases to a certain value then decreases again toward the final state of succession. But such situations can probably be better described in terms of diversity spectra. A spectrum with a plateau is replaced by a steep spectrum. . . . During succession species diversity decreases and pattern diversity increases; that is, diversity and organization shift to higher levels.

The terms *adaptation* and *adjustment* are not used interchangeably throughout this paper. Geographers frequently distinguish between these concepts: *adaptation* means human physiological response; *adjustment* refers to cultural (especially technological) adjustment to an outside or environmental force. *Accommodation* and *adjustment* are used synonymously here.

[3]Population itself may be considered a resource, in which case this statement is a tautology. In this paper, however, population is simply treated as an independent variable.

[4]Millon (1962:80) has concluded that "no simple tendency for centralization of authority over water allocation to increase with the size of the system or the numbers of people involved" occurs in small-scale irrigation societies.

[5]Millon's principal methodological error is that many of his sample's political authority systems are erroneously categorized. Hunt (in this volume) has also noted some problems in Millon's interpretation of "centralization" of authority for several of his ethnographic cases.

[6]Zubrow (1970:11) notes that a "lag" and "overshooting" of population under and over the homeostatic equilibrium point of carrying capacity occurs periodically in all populations. If Kaplan is correct, might building phases correlate with phases of excessive population growth or lag?

[7]Erasmus (1965:277) has employed peasants in Mexico ". . . to excavate and carry earth and rock, to collect data on stone masonry and stone sculpting." These activities were measured to obtain figures on the man-days of labor invested in a cubic meter of fill or masonry and into a square meter of sculptured stone veneer. He computed the "average annual man-day investment. This figure, compared with population density estimates could help fix the number of man-days per year invested by a household. . . . This could be compared with those known for communal projects in societies of varying decrees of organization to estimate the extent of associated political development" (1965:277). To apply Erasmus' computations we need accurate population estimates. Sanders (1968:84) has observed that a great deal of variation exists in speculations about aboriginal populations. The caution against circular reasoning that Kaplan (1963:399) has given for estimating political organization from the size and spectacular nature of monuments applies to estimating population size also: a large population must have been required to build the monuments; the monuments are the mark of the existence of a large population.

[8]The critical points of population growth identified here – e.g. 5 persons per acre and 5000 persons, are not "magic numbers" at which points stress occurs in the system. Rather thay are probably artifacts of the restricted sample that I have chosen.

## REFERENCES


ADAMS, Robert McC.

1965 *The Evolution of Urban Society.* Chicago: Aldine.

1968 Early Civilizations, Subsistence, and Environment. In Yehudi A. Cohen ed., *Man in Adaptation: The Biosocial Background.* Chicago: Aldine.

ALLAND, Alexander

1967 *Evolution and Human Behavior.* New York: Natural History Press.

BEARDSLEY, Richard K., Preston Holden, Alex D. Drieger, Betty J. Meggers, John Rinaldo, and Paul Kutsche

1956 Functional and Evolutionary Implications of Community Patterning. In "Seminars in Archaeology: 1955. Memoirs of the Society for American Archaeology No. 11," ed. Robert Wanchope. *American Antiquity* 22:129-57.

BENNET, John W.

1966 Ecology, Economy, and Society in an Agricultural Region of the Northern Great Plains. In John W. Bennet ed., *Social Research in North American Moisture-Deficient Regions.* New Mexico State Univ. Pp. 35-55.

BOSERUP, Ester
1965 *The Conditions of Agricultural Growth: The Economics of Agrarian Change Under Population Pressure.* Chicago: Aldine.

BOULDING, Kenneth
1962 Where are We Going if Anywhere? A Look at Post Civilization. *Human Organization* 21:162-67.

CARNEIRO, Robert
1967 On the Relationship Between Size of Population and Complexity of Social Organization. *Southwestern Journal of Anthropology* 23:234-43.
1970 A Theory of the Origin of the State. *Science* 109:733-38.

CHILDE, V. Gordon
1962 *Man Makes Himself.* New York: New American Library.

EBERHARD, W.
1958 Oriental Despotism. *American Sociological Review.* 23:446-48.

EISENSTADT, S. N.
1958 The Study of Oriental Despotisms as Systems of Total Power. *Journal of Asian Studies.* 17:435-36.

ERASMUS, Charles
1965 Monument Building: Some Field Experiments. *Southwestern Journal of Anthropology* 21:277-301.

FLANNERY, Kent
1968 Archeological Systems Theory and Early Mesoamerica. In Betty Meggers, ed., *Anthropological Archeology in the Americas.* Washington, D.C.: The Anthropological Society of Washington.

GERTH, H. H. and C. Wright Mills (trans. and eds.)
1958 *From Max Weber: Essays in Sociology.* New York: Oxford Univ. Press.

HARNER, Michael J.
1970 Population Pressure and the Social Evolution of Agriculturists. *Southwestern Journal of Anthropology* 26:67-86.

HARRIS, Marvin
1959 The Economy Has No Surplus? *American Anthropologist* 61:185-99.

HOLE, Frank
1966 Investigating the Origins of Mesopotamian Civilization. *Science* 153:605-11.

JACOBS, Melville and Bernhard Stern
1947 *An Outline of General Anthropology.* New York: Barnes and Noble.

KAPLAN, David
1963 Men, Monuments, and Political Systems. *Southwestern Journal of Anthropology* 19:397-407.

LEACH, E. R.
1959 Hydraulic Society in Ceylon. *Past and Present* 15:2-26.

MARGALEF, Ramón
1968 *Perspectives in Geological Theory.* Chicago: Univ. of Chicago Press.

MAYER-OAKES, William
1960 A Developmental Concept of Pre-Spanish Urbanization in the Valley of Mexico. *Middle American Research Records* 2(8):165-76.

MILLON, Rene
1960a The Beginnings of Teotihuacan. *American Antiquity* 26:1-10.
1960b Structure Within the Pyramid of the Sun. *American Antiquity* 26:371-80.
1962 Variations In Social Responses to the Practice of Irrigation Agriculture. In Richard Woodbury, ed., *"Civilizations in Desert Lands." Univ. of Utah Anthropological Papers No. 62,* December. Salt Lake City: Univ. of Utah Press, pp. 56-88.

MURDOCK, George P.
1963 *Outline of World Cultures.* (3rd ed., rev.) New Haven: Human Relations Area Files.

ORANS, Martin
1966 Surplus. *Human Organization* 25(1):24-32.

PALERM, Angel
1955 The Agricultural Basis of Urban Civilization in Mesoamerica. In Julian Steward ed., *Irrigation Civilizations: A Comparative Study.* Washington D.C.: Pan American Union Social Science Monograph No. 1.

SAHLINS, Marshall
1964 Culture and Environment: The Study of Cultural Ecology. In Sol Tax ed., *Horizons of Anthropology.* Chicago: Aldine.

SANDERS, William T.
1965 *The Cultural Ecology of the Teotihuacan Valley.* Pennsylvania State Univ. Press.

SANDERS, William, and Barbara Price
1965 *Mesoamerica: The Evolution of a Civilization.* New York: Random House.

SMITH, Robert Leo
1966 *Ecology and Field Biology.* New York: Harper and Row.

STEWARD, Julian
1955 *Theory of Culture Change.* Urbana: Univ. of Illinois Press.

WITTFOGEL, Karl
1938 *New Light on Chinese Society.* International Secretatiat, Institute of Pacific Relations.
1957 Chinese Society: An Historical Survey. *Journal of Asian Studies* XVI(3):343-64.
1967 *Oriental Despotism: A Comparative Study of Total Power.* New Haven: Yale Univ. Press.

WOLF, Eric and Angel Palerm
1955 Irrigation in the Old Acolhua Domain, Mexico. *Southwestern Journal of Anthropology* 11:265-81.

WOLF, Eric
1959 *Sons of the Shaking Earth.* Chicago: Univ. of Chicago Press.

WOODBURY, Richard
1960 The Hohokam Canals at Pueblo Grande, Arizona. *American Antiquity* 26:267-70.
1961 A Reappraisal of Hohokam Irrigation. *American Anthropologist* 63:550-58.

ZUBROW, Ezra B. W.
1970 Carrying Capacity and Dynamic Equilibrium in the Prehistoric Southwest (unpublished MS).

CHAPTER 14

# COMMENTS ON THE SYMPOSIUM FROM A HYDROLOGIC PERSPECTIVE

Martin M. Fogel
*Department of Watershed Management, University of Arizona*

It is both a terrifying and rewarding experience for a hydrologist to be invited to a meeting of anthropologists discussing irrigation. It is terrifying in that you begin to wonder if somewhere along the line you missed out on something. You hear terms that are partially familiar such as tidal irrigation, wet-field irrigation, and offtakes. You feel a sense of inadequacy when you hear scientists not of your training use this vaguely familiar terminology to discuss the operation of systems about which you are supposedly an expert.

On the other hand, while slow to take hold, the idea does filter through that these anthropologists do have something to offer in the development of water resources systems. Hence, it is rewarding to a former hard-shelled engineer who has been publicly stating for some time that the major problems of water resources development and operation are not associated with the technical aspects but rather relate to the socio-political situation.

Planners and developers are slowly coming around to professing a need for the social-oriented scientist – the "soft" scientist – to become involved initially in projects concerned with water resources. It makes much more sense to plan for an eventuality than to correct a problem that could have been foreseen.

Having worked in the Middle East, I found the uncovering of past mistakes buried in the sand a fascinating experience. I am reasonably sure that a certain percentage, if not all, of the water-control structures I have designed will meet a similar fate. One such system that will surely soon become part of Middle Eastern archaeology is a system of canals and structures for the delivery of irrigation water to a large oasis in eastern Saudi Arabia. The scheme involved some two to three thousand individuals, some of whom were eagerly awaiting this vital resource. Because of the large number of turnouts in a small area, I decided to use the rotation rather than the demand system for delivering water. The idea was that all the water users had potentially the same requirement and would need water at about the same time. Also, the demand system would probably be too sophisticated for this situation.

The canals and structures were sized based on a rotation schedule. Fortunately, this plan was on paper, and not in the construction phase. Unknown to me at the time was the custom of having water flow by farmers' lands at all times, so that the choice of time of irrigation was theirs. This custom necessitated a continuous flow system, which utilizes smaller canals but is usually more wasteful of water; nevertheless, this local custom had to be considered. A little help from a social scientist in understanding local customs would have prevented a duplication of design effort; much more serious mistakes are made without this bit of consultation.

I was impressed with the calibre and general all-around knowledge about irrigation indicated in the papers. It was stimulating to hear and read about the timing of irrigation water applications, for example. After some one hundred years of agricultural research, it has finally dawned on investigators that maximum crop production may not be the most beneficial or desirable. The same can be true with irrigation. Two or three well-timed water applications may yield greater economic returns to the farmer than a host of irrigations designed for maximum production. Optimal timing of irrigation to conserve a vital renewable resource is an area of research that is receiving the scrutiny of progressive investigators using the modern techniques of computer simulation. Other than in grain crops, results in this field are virtually non-existent at this time. Yet, along comes Downing, who professes no expertise in irrigation but who is able to draw conclusions about the relationship between water scarcity in terms of water availability to plants during their "moisture-sensitive periods" and societal organizations. Downing points out that there has been a failure to refine both sides of the irrigation-society equation. This failure, however, is not only a fault of anthropologists looking into the past, but of water resource specialists planning for the future.

I was a little disturbed by the paper of Adams who discredited the myths concerning the productivity and reliability of irrigated agriculture. Why have we been subsidizing this form of agriculture in the western United States if not for increased production with greater stability? Perhaps the key lies in Adams' statement that for the small farmer, irrigated agriculture is a low-return, high-risk system. On the other hand, an old saying comes to mind about reclamation projects, that is, the development of irrigation projects in the West: it took three generations of settlers to produce one successful irrigation farmer.

Moseley appears to be in partial disagreement with Adams' remarks on the productivity and stability of irrigated agriculture. In discussing early water-management systems in coastal Peru, Moseley states that floodwater farming was not particularly productive. On the other hand, a rise in agricultural output and a major reorganization in residence patterns are associated with the introduction and expansion of canal irrigation systems. In apparent agreement with Moseley, Farrington concludes that there is a strong relationship between irrigation agriculture and settlement pattern on the North Peruvian Coast, but that this relationship is independent of social organization.

While I have no other alternatives to offer, I am not so sure I agree with Vivian that the development of the water-control systems of the Anasazi prehistoric culture in the Southwest was the result of changes in climatic conditions. Such changes are usually very subtle, and minor climatic variations are difficult to detect in modern times as witness the countless discussions on the existence of precipitation cycles. In any case, Vivian goes on to discuss two classes of systems: a conservation-type system where water is used in place, and a diversion-type system. The point is, if we had studied and learned more about either one of these systems, the heartland of the United States may not have experienced such a miserable time as it did during the "dirty thirties."

Neely presents information on another part of the world: the qanats, dry farming terraces, and check dams that existed in Persia some 2500 years ago. He states the interesting hypothesis that population growth, the introduction of water-control and irrigation systems, and the changing settlement and community site patterns, are a direct result of a planned expansion program promoted by the existing government. Thus, our water resources activities in relation to settling the semiarid Southwest may be viewed as an example of planned growth. One wonders, however, just how far this "planned expansion" should go when parts of southern California and central Arizona are considered.

In an effort to predict sociological differences, Spooner studied the technological systems of two small-scale irrigation societies in Persia. These differences, concluded Spooner, could be predicted from ecologically oriented studies that illustrates the complexity of the irrigation system. As one looks at some of our older irrigated areas and views the varied patterns made by the many systems of irrigation and drainage canels, it is easy to understand how this variable can be important in analyzing sociological systems.

Dealing with the Near East, Gibson rejects the existence of a close relationship between a flourishing irrigated agriculture and a stable and vigorous central government. Using evidence from Mesopotamia, he proposes the contrary, that intervention by state government tends to weaken the agricultural basis of a country. What is of greater importance to a hydrologist, however, is the reliance he and other anthropologists have to place on understanding such complexities as sedimentation, salinization, and plant-soil-water relationships, in order to understand some of the social and cultural aspects of the existing society. With the Euphrates River at a higher elevation than much of the irrigated area, and poor irrigation practices contributing to the high water table, there may have been a lack of understanding of the interrelationships of groundwater movement, drainage, and salinity. Nonetheless, fallowing in alternate years was utilized as a successful solution to the problem. Unfortunately, water resources specialists and engineers of today have not learned their lessons from the past, as witness the current situation in the vast Indus River Basin of Pakistan. About one-fourth of the 30 million irrigated acres is encountering serious drainage and salinity problems. One can only begin to speculate on the social and political implications of such a situation.

Another interesting point was made by the Hunts, who suggest that when water becomes scarce, that is, when there are population pressures on land and water resources, a situation arises under which centralization of authority is adaptive to reducing conflict. I can recall a conflict under a different set of circumstances, an overabundance of water, which involved the alignment of a drainage ditch. Simply stated, the problem was that one group of people wanted the ditch to cross a county line while another group didn't. The conflict grew so intense that at one particular attempt at mediation, I was left standing between two lines of shotguns. Fortunately, the conflict was resolved when it was decided to direct the unwanted drainage water towards a neutral zone, a solution which was not necessarily the most economical.

Kappel's goal was to demonstrate mutual relationships between the three variables – population growth, extent of irrigation systems, and centralization of political decision making – using various world cultures. The conclusions appear to support earlier studies in that major changes in size of the irrigation system occur only after changes toward centralization have been made. This sequence seems to imply that the development of irrigation projects takes place more rapidly under a more centralized form of government, perhaps to its own detriment.

Performing as an anthropological systems analyst, Lees states that the function of canal irrigation should be viewed not as constant, but as variable in time and place. She goes on to say that the installation of new irrigation devices whose construction and operation is often beyond the comprehension of a local peasant community, such as found in Mexico, causes the subsystem or local community to become more responsive to stimuli or perturbations of the state government, the system. She feels that this reduction in local autonomy is potentially unstable.

The development of irrigation from groundwater sources in many parts of the world, including northern Mexico, has eventually resulted in a certain amount of instability. Extending the irrigated area beyond that for which water is available has resulted in serious consequences. Not only are water tables lowering, but what is worse, the remaining groundwater supply is becoming contaminated with sea water intrusion. Lees does not distinguish whether the resulting instability is due to the introduction of technology itself or to the lack of government intervention. Failure to completely understand the physical system is a major cause of this instability in many current situations, although political motivations also contribute to instability.

Water control in the terraced rice lands of southern and eastern Asia is the subject of Spencer's contribution. The term *water control* is substituted for the word *irrigation* and perhaps rightly so, as water flow into and out of the system of terraces much be properly regulated. It is interesting to note that the development of the terraced field groups requires decades and even centuries. The time needed to activate currently proposed water-resource projects may approach such a duration not because of complexity in design and construction, but rather the aforementioned socio-political aspects of the situation. A prime case in point is the diverting of water from such river basins as the Columbia and Snake into the Colorado. The odds are very great that this debate, barely in its infancy, will rage on for many years to come.

The practice of mountain irrigation in the Swiss Alps as presented by Netting is relatively unheard of. However, who can argue with the fact that hay production is increased four to five times? Perhaps one day the downstream water users, such as the municipalities and the recreationists, will raise an outcry to put the water to a more beneficial use.

In summary, I was continually amazed at the high degree of understanding the contributors had of the physical systems. To an engineering hydrologist, it was a rewarding experience that suggests the utility of close cooperation between the physical and social sciences.

CHAPTER 15

# COMMENTS ON THE SYMPOSIUM FROM A GEOGRAPHIC PERSPECTIVE

Chris Field
*Department of Geography, University of Montana*

Geographers have not usually been interested in broad propositions such as are implicit in the title of this symposium. I know of little relevant literature which might effect a cross-disciplinary transformation. Since geographers have largely regretted their earlier fling at environmental determinism (how does environment shape society and civilization?), most have tended to avoid broad commitment to questions which assume that single variables provide causes to which social changes are the effects. Having also long been concerned with explanation of the distribution of variation in physical nature as well as with the differences in human use of that habitat diversity, geographers are reluctant to embrace any explanation of societal change or human behavior which postulates some single systemic externality or environmental condition as a sole or prime cause. Just as the explanations of social condition or change as the result of environmental "demands" or "conditions" prove unsatisfying, so are the reasonings that would root sequences of societal events or modes of developmental elaboration in the demands or natural consequences of particular technologies.

The idea that irrigation or any other element of a set of technological skills or environmental conditions should be treated as a single variable may be useful for the purpose of analytic description of particular societal operations. But, in today's understanding of the multiplicity of human relationships to their habitat, a single element, causal, and linear approach to a search for general principles seems less fruitful than ever. We need conceptual vehicles based on continuing processes which link societal changes within the events and systemic relationships that flow and interact over time. Simple stimulus-and-response models no longer afford satisfactory explanations, as is evident from the kinds of questions raised here.

The papers of this symposium, however common their focus of interest, seem to share no clearly perceived process which could link all the evidence of societal change and irrigation development. It is apparent that some of the papers rise to contest the universality of Wittfogel's generalizations and speculations. In a way, this collection serves as a tribute to the power of a challenging hypothesis. But, the evidence and arguments presented here make it unlikely that we will ever find universality of developmental causation in association with irrigation and type of society, Oriental Despotism, or not, unless that type of society be defined largely on the basis of its dependence on massive "large-scale" irrigation.

The notion of societal adaptation, retrospectively perceived as an accommodation to environmental conditions or to systemic alterations within the boundaries of any social unit, may also prove to have low general explanatory or predictive power. The obvious remark for a geographer to make, is, that since there have been settled places in which irrigation would have been "adaptive" but was not adopted (i.e., aboriginal California), it becomes inconsistent to say the techniques of irrigation are, in principle, responses to either necessities or environment. Since this symposium clearly demonstrates that many sorts of social organization at various scales, both past and present, depend on irrigation, it is no longer plausible to assert irrigation technology as the major source of any particular style of government or degree of "adaptiveness" of a society. It appears to be too easy to drift into either exceptionalism or the vague generalities of cultural relativism into which we project our own urban values and convictions about authority. One must also be wary of the "adaptation" reasoning, since it may lead to either environmental, technological, or a tautological "social system" kind of determinism. Unless process mechanisms are sought and their operations understood, neither hypotheses or general principles will bear scientific scrutiny.

Most of the models of societal development and change with which I am familiar seem to depend on internal developmental self-stimulation, after some external initiating situation or event. Do we really think that human society is able to maintain itself as an exclusively closed system of operation? The conception of systematic self-perpetuating adaptation and development seems to presuppose the inevitability of change towards increases in numbers and elaboration of societal organization. Such a system would require ever increasing inputs of free (low entropy) energy, obtained at the price of increasing efficiency of diversion for transfer to man. If this be true, then environmental disorder is the only possible result. It is foolish to believe that human ingenuity could forever manage such a thermodynamic feat. Drastic change to a living system would come because of growth to (or, inertially, beyond) a threshold at which the fundamental biotic resources had been simplified, extracted, eroded, contaminated, or otherwise degraded so that under known (feasible) technologies, the usable production declines below all perceived margins of utility. The food surpluses at the lower level are then simply not available for transfer into the more elaborate specialized structures of social organization.

That is why this geographer is convinced that much attention should be given to the environment-changing capacities of societal organization, including technologies such as irrigation as pervasive integrating factors. Even a deterministic view of societal development based on the processes of ecosystem simplification, diversion, and disruption might still provide valuable insights. It certainly appears that many distinctive past civilizations have either exacerbated or provoked serious environmental disruptions at the start of their decline. There is a distinct possibility that we are consciously, yet somehow irresistibly, on the same slide.

Irrigation *per se* does not cause population growth, despotic government, soil conservation, or resource destruction. It can be, and in fact, has been one of the more stable technologies of agricultural production. Provided the local flows of water and nutrients remain in roughly the same proportion as those in the pre-agricultural economy, irrigation agriculture can approximate a steady-state system. (See Spencer, this volume.) One may interpret the human history as one of invasion, interference in succession, and eventually, successful dominance. The irrigated wet-field terraces certainly are a specialized anthropic ecologic niche. But, the language of ecological concepts fails us when it comes to understanding social change. The process of innovation and the rates of cultural transfer operate differently in society; the successions seem too abrupt, disorderly, and unpredictable.

One important element of change in irrigation-dependent systems, once they have expanded to the perceived limits of land space or water supply, is intensification. The techniques of intensification normally fall under what we commonly term "sophisticated" technology, with all the context relativity the situation demands. They include terrace building, double, triple and intercropping, plant selection, fertilization, or even monocultural concentration on a particular crop. The changes are designed to make more efficient use of space, time, water, or biological material, not necessarily labor. Many of the intensification techniques represent genuine innovation. But, when improvisation ceases, innovation fails, and external information is not available, the limits of known technology are reached. The localized system will probably stress its boundaries. As intensification runs its course, before, or maybe after, these stresses are perceived, the social organization will probably produce some effort to enlarge. Should this effort involve incorporation (within the boundaries) of improvable or substitutable resources of another place, we read the results as conquest, political aggression, imperialism, or "systemic response" to extra-local opportunity. Territorial expansion may provide an opportunity to repeat the process, but eventually the physical limits of habitat are encountered. Should the stresses of the effort to maintain organization continuity (including production and consumption habits) take the form of further intensification regardless of the declining marginal returns, some relatively permanent damage to the productive land is inevitable. Intensification carried to the point of destruction of the biotic network underpinning any food production system will force some kind of social change. I think it more likely to be regressive, since I know of few instances of fundamental innovation being produced in the context of dire necessity.

On earth, expansion of any kind inevitably leads to an encounter with environmental limits. The encounter may be postponed over time and space by technical or organizational innovations accompanied by ever increasing and less rewarding inputs of energy. But as long as the circulation within the system is dedicated to maximizing growth by positive response to the externalization of profit within the productive subsystems, the resulting habitat simplification for thermodynamic efficiency will produce a negative environmental control. The price of material intensification, whether the goods be social or subsistence, is, as Lees suggests, the ever more unstable organizational response to the newly seen and encountered system boundaries. By increasing response to extra-local conditions, the effective communication response time to all indicators of local environmental limits falls. The growing basic production system becomes less responsive, less stable, and more prone to major environmental disaster. I suspect that "despotism" or eventually anarchy, as a type of societal structure, could arise from desperate attempts to maintain growth rates, or institutional role-bound privileges. It may also represent an attempt to hold together an increasingly precarious administrative structure within which the dependencies have shifted from the localized, internally diverse sets of ecological relationships with food production, to the broader scale relationships of specialized institutions within which less substitution is possible. (The Pentagon is not likely to be able to shift individually or collectively to the task of potato growing, irrigated or not. But the local chief or priest is more likely to be able to farm or go hunting in crisis time.) Societal institutions, with increasing size produced by multiplication of numbers and intense specialization, may have built-in resistances to disintensification. The inertial aspects of this resistance, both cultural and in information flow, is a characteristic of bureaucracies.

As a species, by virtue of the lack of world-wide historico-cultural uniformity, we have survived despotisms and bureaucracies. Their downfall has practically always been the failure of growth, and, of course by definition, the decline starts at the apogee. We have reached, in the past, the relative limits of intensification of technique. Seeing this, we seem to have proclaimed that next time, being limited only by human creativity, technological intensification has no limits. It appears that the most environmentally significant over-applications of techniques follow from conditions of abundance, and are applied by social structures independent of local habitat variability and too specialized to recognize or accommodate fundamental biological realities. There are limits to our interference with terrestrial energy flow. We are closer than ever to them today.

Irrigation, like any other technique, must be scaled in relationship to the resources of the place. (The N.A.W.A.P.A. scheme is one of the grandiose diversion plans which takes no serious accounting of local environmental change.)

If the functions of the production system are not directed towards a relatively steady-state economy in an increasingly diverse and intricate biotic community, the energy which should circulate may either escape more efficiently or accumulate to some disaster potential. Great population fluctuations, floods, even desiccation are typical natural consequences.

Irrigation is one of the production-enhancing techniques which has provided the food to make it possible to experiment with a range of social organizations. If it is true that men experiment and change as a consequence of plenty instead of necessity, irrigation is one singularly important element in the geography of social opportunity. That some have created the small, yet often extensive, stable developments, while other irrigators have founded and lost empires, can be taken as the basis for hope that we may yet have a chance to learn how to survive in a world made safe by and for diversity. The lesson should be obvious.

The following general response was mostly stimulated by Gibson, Lees, and Adams. We should, by now, be able to marshall some hindsight over our collective experience with irrigation. The most immediate applicability of this experience might well be applied to examination of some contemporary governments – their administration, megalomania, and response to pressures for "economic development." The way technology is used, in development and intensification, either reflects tacitly accepted social purposes or indicates the more-or-less conscious governmental purposes. Following the Boserup model, Gibson points out a probable interpretation of the decline of some of the Mesopotamian irrigations. Is it the structure of big government which generates the demands that cause the population growth system to override the common sense and abilities of the working farmer? At what scale of enterprise, be it irrigation, manufacturing, trade, or government, do localized diversities of social, ideologic, and economic structures literally get engineered out of existence as their environmental bases are inalterably changed?

Developments which indicate the importance of this kind of change may easily be seen today in Spain. Over at least two millennia, the Iberian peninsula has been exposed to the irrigation technology of the Mediterranean and Near East. In the present landscape, through Andalucía to Valencia, all the Mediterranean techniques survive. The Islamic period seems to have been one of great experimentation and development, expanding on the remnants of the largely urban Roman developments. Qanats, flood and silt trap terraces, tanks, and sub-alluvial dams all function from that period. More significant, perhaps, is the survival of egalitarian Islamic institutions of water control, most of them based on management of waters brought by canals from a variety of basically surface sources. Until recently the majority of communities of irrigators in eastern and southern Spain had a high degree of cohesiveness, albeit frequent internal conflict, but with independence from regional government. Water sources were one significant element in that cohesiveness. Among other trends today, such as outmigration to industrial employment in northern Europe, there exists an invasion by new technology based on pipes, pumps, well drilling, and political power. The spatial-locational factors of traditional irrigation communities have been upset. In arid Spain, the value of land traditionally lies in its irrigability, with rights to water. The new technology has an urban source and an international commercial context. As the new possibilities have been exploited, limited only by the law requiring wells to be more than one hundred meters apart, traditional surface sources of water have declined. Cheaper interfluve and previously unirrigable land is bought or leased by capital not available or controlled locally. The old allocative institutions have been forced to cope with declining quantities and increasing conflict. The fact of social disruption is seen, it appears, as a tool of economic development. The communities are forced into a national dependency relationship as one by one they feel forced to appeal to the central government for inclusion in one of the major irrigation development projects. The old lines of local power change, as do the economic dependencies and the decisions of production, planting, timing, and quality. The best lands often change hands or suffer a decline in productivity. The traditional farming communities are being forced off their lands and out of their societal relationships into an industrial economy. The exploitative pressures are so great and governmental awareness so dim, that even today, salinization is deliberately committed. (Leased land is irrigated from saline wells at Carboneras for a planned maximum of four years of melon production.) Technological change of irrigation is clearly an instrument in this process. How deliberate this exploitation is must rest in the consciousnesses of the planners, technocrats, and political decision-makers. Documentation may be hard to find. The changes now occuring may well be analogous to those we try to reconstruct from the sherds and landscapes of ancient times.

Although the following comments on Lees' paper were not all expressed at the symposium or during the agreeable discussions before and after, they represent a selection from my notes of what then came to mind. The language of systems is appropriate to the complexity of hydrology and society. Combined with a no-end, no-beginning kind of ecologic framework to thought, emphasis could profitably fall on processes.

Is there any way for "cohesiveness" in organization and reduction in societal stability to be found analogous to the ecological stability that is thought to result from diversity of speciation and niche exploitation? The sheer complexity of a diverse biotic system is thought to maintain stability of population numbers. Is what we see as societal stratification and organizational elaboration really a comparable diversity? I think not. Whatever the scale of energy, material, or information flow in a definable human system, does the cohesiveness, centralization, despotism, or monolithic authoritarian style become a pre-condition of some de-stabilizing change? If this possibility really exists, and I think it may, I am still waiting to understand the actual process by which irrigation can bias the system into an authoritarian or even a regionally integrated mode. Extra-local response capacity may well ultimately be reduced by institutional inertia, especially if the system dependence on massive non-localized environmental changes is high (conurbations, national transport systems, inter-basin water transfers, nuclear powered energy grids, etc.). I cannot disagree with the basic plausibility of Lees' argument. I, too, shall gloss over the cyclical assertions since I know too little about information flow and glut to assert it clearly enough as a fundamental process in the system. I agree with the conclusion that organizational change occurs with technological change, but I do not see it simply as a "byproduct."

With respect to the facts of the Oaxaca irrigation schemes, I am not well informed, but feel that nationally, the Secretaría de Recursos Hidráulicos has concentrated on "large" projects. (Ing. Adolfo Orive Alba, long head of Recursos Hidráulicos, defensively reacted at length to what he felt was a criticism – the lack of small irrigation works. In Anexo No. 6 in *Politica de Irrigación en Mexico,* Orive Alba reports a total of 1,540,039 Ha. in large projects against 359,353 Ha. in small ones over the years of 1926 to 1958. In a not exactly consistent defense he also states that 2,366 [undefined] *pequeñas obras,* irrigating 361,353 Ha. were built. He goes on to assert, among other things, a greater cost effectiveness for large projects.) From the point of view of the peasant, the large projects represent almost the ultimate form of governmental dependency.

I fail to see how Lees can postulate that the "progress" system eventually leads to a "a new equilibrium" wherein "the system may begin to respond once again to local conditions." If that response is to the habitat deterioration or to the limits of resource supply within technical context, how can it be stabilizing? Perhaps the word "retro-stabilization" would be good! The only recourse, it would seem, is to re-define and enlarge the boundaries of the system. Growth is the response to growth? In spite of the above questions, I feel that the major conclusion is true; hyper-dependence makes large administrative systems unstable.

Specific historic and ethnographic comparison seem to invite an easy recourse to uniqueness and chance by way of explanation. Spooner has avoided that danger and has managed to draw some very intriguing insights out of what must have been both exciting and frustrating field work. His attempts to clarify definitions in typology and subsystems are very useful, because there has been a tendency to assume a lot of common understanding. His remark about qualitative distinctions being significant rings very true, even when attempting to deal with the very difficult problem of definition of scale, as Kappel does. Spooner is also on the mark when he reminds us that basic typologies or scheme(s) of definitions must be subservient to the generative model.

When it comes to the problems of scale, the geographer should be quick to point out that irrigation systems not only integrate man, water, and land over time, but also over space. The spatial dimensions of scale in irrigation systems seem to be hinted at; I suppose that is what is meant by local and small-scale. Unity in resource quality, whether it be amount of rainfall, size of streams, or kinds of soils, has often been the basis for recognition of small regions which consist of basically similar and related topographic units. These units may in some useful way be related to the human occupance or some kind of cultural unity, be it economic, political, or even linguistic. Relationships between the scales of locality and regionality are most elastic, stretching to fit the conceptual scheme of immediate interest. The simplest and often imprecise habit of thought is that the valley, set of hillslopes, piedmont, plateau, or other topographic unit, is the unit of small scale. For irrigation one must involve the definition of drainage basins, the smallest and uppermost of which is called the first order, followed by mergers into the second and so on downstream. In an area of great relief or considerable variety such as the Valley of Mexico, the concept will not help the anthropologist's scale problem very much. An arbitrary conventional spatial scale limitation to second order basins might help minds meet, but it could also produce a rash of examples of interbasin transfers performed, as they often are, at a rather elementary technical and organizational level. In suggesting that space and physiography are not the final solution to the scale problem, I do not mean to derogate the significance of physical factors such as flow rates, soil permeabilities, and distance/time ratios from source to cropland. These are factors that may well be technically made more efficient by intensified and more organized efforts to allocate the resources. In connection with Spooner's perception of the investment subsystem and the mutually reinforcing process of accretion of capital and control, isn't is possible that the organization of metallurgical technology might well have functioned similarly?

An excellent review and valuable analysis, Kappel's work ought to be elaborated, perhaps in matrices of some kind. If there are significant bumps, groupings, or plateaus in our continuities of scale, in the social organization of irrigation, it would be good to find more of them. The heterogeneity of data Kappel fitted to this scheme ought to balance out the likelihood that the categories are simply our own cultural projections. The business of defining large and small scale is a perplexing one, all the more since typologies, assumptions, and some rebuffs to Wittfogel rest on the distinction.

If, as Spooner has begun, an acceptable typology could be fitted into the enormous range of ethnological and archaeological observation and inference, some progress might be made towards identification of critical variables limiting developmental processes of civilization. With no meeting of minds over the scale of generalization, principles applicable to several levels will be more difficult to extract.

Geographers have been working over the phenomena of settlement for years, but have no sure methodology or rank-size discrimination generalizations that seem immediately useful here. Since most of our quantifiers have depended on someone else's data, they have been content with census districts, townships, counties, and chambers of commerce. There is a literature about "journey to work" and time/distance behavior, but it is largely industrial, without easy applicability to proving association of hierarchies of settlement size and location with agricultural work sites. We need to know more about these behaviors under various technologies.

If it is indeed true that social investment in centralization of management does come before elaboration or intensification of irrigation technology in a developmental growth sequence, a whole series of economic ideas, from capital formation to social consumption may change. The scalar shift Kappel points to at populations of 5000 may be very significant in relation to the concepts of role specialization underlying urbanization. Linked to theory in communication and administrative effectiveness, some principles of scale in social organization may be discoverable. I intuitively feel that scale and coherence of social functioning are interdependent in relation to neighborhood, community, and city size. My intuitive feel for the organizational function of irrigation systems is primarily related to scale in connection with sophistication of technology, size of flows, extent of area, and efficiency of communication within the production subsystem.

Downing's contribution is valuable for the evidence, often missed by urban-oriented speculators in societal change, that environmental and production systems realities exist. That one can bring modern "scientific" agronomic knowledge to a confirmation of the practices and understandings of peasant farmers is not really news, but it is comforting to those who wish to see agricultural systems as environmentally adaptive. If two systems of thought converge, can reality be far away? In irrigation, as in any other livelihood, men cope (eschew adaptation?) with their understandings of these realities by integrating their knowledge of technique and perception of the possible with experience. Physical uncertainties have to be encountered within the structure of the possibilities that are known or thought to exist, whether the perceivers be farmers or bureaucrats. Lacking a farmer's understanding of crop needs, natural variation in water or land distributions, there should be little surprise that large, extra-local and massive irrigation schemes eventually fail. If there is a great informational gap between locality and water-management authority, whether that gap be technical, in the agronomic sense, or ideological, it would seem inevitable that production or the continuity of the resource base should suffer.

This coherent account of exemplary local wisdom in flexible adoption of social concepts and mechanism to regulate water is unusual. All the Díaz Ordaz parts fit. The egalitarian bias of water allocation could have come, as well, from Spain, with the language and similar, but far from identical roles. Variations on similar local management themes may also be found in the hill country of eastern Andalucía.

That hypothesis relating critical moisture-sensitive periods to some kind of forcing of rigid allocation system seems reasonable, given some of our own cultural preferences for equability. Downing is absolutely right that the ideological nature of the allocation is a cultural matter. In a larger sense, however, so is the notion that allocation should reduce conflict. Tension and conflict, whatever its organizational resolution of the moment, may be more closely and permanently linked to population growth potential than to the divisibility of the hydrologic flow, hunting territory, or other resource.

Moseley's speculative reconstructions sound plausible enough. I am inclined to agree with the steeper terrain hypothesis – even to the point of suggesting that the staple food argument may well apply because of introduction from the highlands of both valuable cultigens (tubers in addition to maize??) and diversion with distribution of water. Evidence for very early lowland floodwater farming and small diversions for irrigation will be as difficult to find as evidence for earlier highland channel diversion, simply because the same favorable sites have long been in use.

Given the topography of the Peruvian valleys mentioned, the argument for earlier use of the upper portions of the coastal piedmont for steeper gradient canal irrigation makes good sense.

My principal concern is for a process that could link the "demographically precipitated commitment" to "more than casual farming." It seems that some kind of Malthusian necessity lurks in the background! Does it always stimulate an "adaptive" innovation? Surely there must be some failures for archaeologists to find?

# INDEX